Efficient Methods for Valuing in Mathematics Education

Efficient Methods for Valuing in Mathematics Education

Edited by
Tim Hopkins

MURPHY & MOORE
www.murphy-moorepublishing.com

Published by Murphy & Moore Publishing,
1 Rockefeller Plaza,
New York City, NY 10020, USA

ISBN: 978-1-63987-180-3

Cataloging-in-Publication Data

Efficient methods for valuing in mathematics education / edited by Tim Hopkins.
p. cm.
Includes bibliographical references and index.
ISBN 978-1-63987-180-3
1. Mathematics--Study and teaching. 2. Mathematics. 3. Mathematics--Study and teaching--Evaluation. 4. Education. I. Hopkins, Tim.

QA11.2 .E34 2022
510.71--dc23

For information on all Murphy & Moore Publications
visit our website at www.murphy-moorepublishing.com

MURPHY & MOORE

Contents

Preface

The practice of teaching and learning mathematics in contemporary education, with its scholarly research is termed as mathematics education. The study of whole numbers and their arithmetic which encompasses addition, subtraction, multiplication and division, and fractions and proportionality takes place within this field. Some of the efficient methods of teaching mathematics are computer-based mathematics education, computer-based math, conventional approach, exercises, problem-solving, etc. In computer-based mathematics method, computers are used to teach mathematics whereas in computer-based math method, mathematical software is used as a primary tool of computation. This book elucidates the concepts and innovative models around prospective developments with respect to mathematics education. It strives to provide a fair idea about this discipline and to help develop a better understanding of the latest advances within this field. Those in search of information to further their knowledge will be greatly assisted by this book.

Various studies have approached the subject by analyzing it with a single perspective, but the present book provides diverse methodologies and techniques to address this field. This book contains theories and applications needed for understanding the subject from different perspectives. The aim is to keep the readers informed about the progresses in the field; therefore, the contributions were carefully examined to compile novel researches by specialists from across the globe.

Indeed, the job of the editor is the most crucial and challenging in compiling all chapters into a single book. In the end, I would extend my sincere thanks to the chapter authors for their profound work. I am also thankful for the support provided by my family and colleagues during the compilation of this book.

Editor

1

Student and/or Teacher Valuing in Mathematics Classrooms: Where are we Now, and Where Should We Go?

Monica E. Carr

Abstract A seminal literature review of values in mathematics education was conducted at the turn of the century, and at that time revealed a paucity of research in this area (Bishop et al. in Values in mathematics teaching: The hidden persuaders? Dordrecht, The Netherlands: Kluwer Academic Publishers, 2003). Bishop and colleagues noted that a change in the values being taught is implicit in any recommendation for changing teaching, and argued that any significant development in mathematics education probably implies a change in values. Research in values in mathematics education remains a high priority today as STEM participation and achievement around the globe continues to encounter many challenges. This chapter presents an updated systematic literature review of values in mathematics classrooms with a view to identifying what has been achieved more recently in this field. Using a systematic search of peer-reviewed publications, some 299 abstracts met key term search criteria. Following an examination of the abstracts, a final data set of 34 studies were retained for further review and analysis. Research methodology, geographic location, stakeholder—teacher or student—valuing, age, grade level, gender, and a summary of original main conclusions were reported for each of the relevant studies. Results were synthesized across the data set to describe where the body of research is at currently.

Keywords Literature review · Mathematics · Students · Teachers · Values

Introduction

The study of values spans a broad multi-disciplinary terrain, with different disciplines pursuing the central concept of values from unique orientations. In the seminal literature on values in anthropology, Kluckhohn and Strodtbeck (1961) wrote that values can be conceptualized as being able to answer basic existential questions and to help

M. E. Carr (✉)
Melbourne Graduate School of Education, The University of Melbourne, Melbourne, VIC, Australia
e-mail: mcarr.phd@gmail.com

provide meaning in people's lives. To social scientists, values are viewed as a means to help ease conflicts between collective and individual interests. Through this lens, values can be conceptualized as serving an important function in which individuals can work together to attain goals that are ascribed to by the collective group (Parsons and Shils 1951).

Although many definitions of values abound in the broader literature, unique meaning and role of values and valuing have been defined by the mathematics teaching and learning community. In his seminal discussion on culture and values in mathematics, Bishop (1988a) introduced six fundamental activities that he argued are universal, necessary and sufficient for the development of mathematic knowledge: *counting, locating, measuring, designing, playing,* and *explaining.* Subsequently Bishop (1988b) envisaged values as a variable of affect, and went on to describe six values that underpin the widely utilised notion of Western mathematics: *rationalism* and *objectism; mystery* and *openness; control* and *progress.*

At the start of the 21st Century the first literature review of values in mathematics education was conducted by Bishop et al. (2003). Reporting a dearth of literature, Bishop and colleagues noted that most empirical research was conducted within the five years prior to their review. In particular, the ARC Project *Values in Mathematics Project—VAMP* (1999–2002) was awarded to Bishop and Clarkson, whilst around the same time a parallel project led by Lin and colleagues was conducted in Tawain. Bishop and Clarkson reported that some studies covered in their literature analysed the values portrayed by text materials used in teaching mathematics. Other studies focused on mathematics classroom teachers, and on values-related activities within the classroom.

Bishop and colleagues (2003) attributed two main reasons for the paucity of research at the intersection of mathematics education and the values area. Firstly, the universalism of mathematics, in which mathematical concepts transcend language or geographic location together with the universal applicability of mathematical ideas, fosters the belief that mathematics is culture-free and therefore value-free knowledge. Bishop and colleagues explained that this universalism is one of the prime values underlying the "western" notion of mathematics that has gained pre-eminence in all parts of the world. Secondly, Bishop and colleagues described the long-held belief that mathematics teachers do not need to take social aspects of mathematics education into account in their teaching, which has resulted in the technique-oriented curriculum in which skill teaching and learning are the central focus. Bishop and colleagues reported that any significant development in mathematics education probably requires challenging these established beliefs. Accordingly, Bishop and colleagues argued the importance of taking values into consideration in future mathematics education research emphasizing that a change in the values being taught is implicit in any recommendation for changing teaching.

Initially, Seah and Bishop (2001) defined values in mathematics education as:

> ... One's internalisation, cognitisation, and decontextualisation of affect variables (such as beliefs and attitudes) in one's socio-cultural context. They are inculcated through the nature of mathematics and through one's experience in one's socio-cultural environment and in the mathematics classroom. These values form part of one's ongoing developing personal

> value system, which equips one with a pair of cognitive and affective lenses to shape and modify ones way of perceiving and interpreting the world, and to guide ones choice of course of action. They also influence the development of one's other beliefs and ones needs in mathematics education and in life (p. 444).

Definitions of values in mathematics education have continually evolved since that time, with Seah (2018) most recently writing that:

> ... valuing refers to an individual's embrace of convictions which are considered to be of importance and worth. It provides the individual with the will and grit to maintain any 'I want to' mindset in the learning and teaching of mathematics. In the process, this conative variable shapes the manner in which the individual's reasoning, emotions and actions relating to mathematics pedagogy develop and establish (Seah 2018, p. 575).

In earlier literature, values were viewed by Bishop as an affective variable. An important distinction between this and the current definition proposed by Seah is that values are viewed as a conative variable. In light of global diversity driven by modern migratory trends, in which students, teachers, and parents are submersed in new cultures, an "individual's embrace of convictions" is arguably of particular significance.

Science, Technology, Engineering, and Mathematics (STEM) form the backbone of our current global economies, with sectors such as education, engineering, food production, health care, infrastructure, manufacturing, research and development, supply chain, and transportation relying heavily on a STEM skilled workforce. Arguably, achievement in mathematics is vital to the adequate preparation of students to meet the technical needs of jobs of the future. However, falling rates of participation and achievement in STEM subjects has been widely acknowledged in Australia (Timms et al. 2018). Seah (2018) has highlighted the significant, though often overlooked, role of values in supporting the cognitive development and affective state of mathematics students. Accordingly, research in values in mathematics education remains a high priority. Common to any exploration of values, some form of measurement of the values held is necessary. Value measurement requires a questioning process through which themes are explored such as: what values are held by individuals?; how are various values prioritized?; and what variations or similarities in values may exist amongst cultures? are explored. In response to these challenges, and shaped by these value measuring aims, *The Third Wave Project* led by Seah commenced in 2009. At the time of this writing nearly all active researchers in the field of values in mathematics education have been invited to participate.

Primary sources of literature provide first-hand information on studies and includes detailed descriptions of the studies' methodology, data, analysis, results and findings. Although published in a variety of sources including journal articles, book chapters, dissertations, or conference papers, a review including all primary literature sources is beyond the scope of this chapter. As such, this current review has been restricted to peer-reviewed journal articles that arguably reflect the most current and complete studies that have undergone a rigorous review process. Presenting an updated systematic literature review of values in mathematics education, this study aims to provide a map of the empirical research conducted to date, and to assist future researchers shape their exploration in values and valuing in mathematics education. The following research questions were developed:

i. Where has research been conducted?
ii. Which stakeholders are represented in the research?
iii. What is known about the development of values?
iv. How consistent are the findings reported in the studies?

Systematic Search Procedure

The What Works Clearinghouse (WWC) Procedures Handbook Version 4.0 has been developed by the U.S. Department of Education's Institute of Education Science (IES) to facilitate a systematic literature review process that uses consistent, objective, and transparent standards and procedures (WWC Procedures Handbook V4.0, 2017). The WWC review process comprises five steps: developing the review protocol to define the parameters for the research to be included in the review; identification of relevant literature; screening studies; reviewing studies; reporting on findings. While values research in mathematics education is a relatively young field, the WWC systematic review framework was adopted for this current chapter to provide a replicable procedure for future researchers working in this field.

Studies were located by conducting a systematic search of peer-reviewed literature published between January, 2003 and March, 2018. Both the PsycINFO and ERIC databases were queried using the search terms *math** AND *valu**. The abstract of each article was examined to determine whether an article was likely to meet inclusion criteria for further review, and a review of the full article was conducted when further clarification was necessary. Inclusion criteria required that:

1. The study reported on mathematics "teaching" or "learning" for students studying at primary or secondary levels
2. The study reported empirical data that may have been gathered from: classroom work; project work; homework; assessments; classroom observations; field notes
3. The study reported on either teacher, student, or parent/guardian valuing
4. The study investigated values alone, or in conjunction with other components of mathematics education such as test anxiety, personality, and/or beliefs
5. The full article was published in English in a peer-reviewed journal.

The psycINFO database search identified 109 abstracts that met search term criteria. Following examination of each abstract, 67 articles were retrieved for further clarification. Of the 52 studies that met the inclusion criteria (see 5 points above) for this review, 21 were published before 2003 and thus omitted from further review. Two journal articles—that is, Dede (2006) and Eklof (2007)—were unable to be located and were subsequently omitted from further review.

The ERIC database search identified 190 abstracts that met search term criteria. Following examination of each abstract, 16 articles were retrieved for further clarification to determine adherence to inclusion criteria as the abstract alone provided insufficient information about the study. Three studies from the USA, one study covering both the UK and Canada, one study from Malaysia, and one study from Africa studies did not include empirical data and thus were omitted from further review. One study from Taiwan provided data for science rather than mathematics, and was also omitted from further review. One study that explored values in Hawaii was unable to be located either in the university library or elsewhere (Furuto 2014), and was omitted from further review.

Results and Discussion

The search and study inclusion procedure identified 34 studies that reported on a variety of stakeholder perspectives as they relate to values and valuing in mathematics education. The Appendix provides a descriptive overview of each study included in the review. The number of annual publications were plotted in the line graph depicted in Fig. The trend line indicates a positive growth in publication volume, reflecting the growing interest by researchers in this field.

Where Has Research Been Conducted?

Studies were conducted by 30 research teams, of which two studies are affiliated with the Third Wave Project (Dede 2013a, b). The studies presented in this review were conducted in 14 countries: *Australia (1), Canada (3), Finland (1), Germany (12), Greece (2), Hong Kong (1), Israel (1), Norway (2), Singapore (1), South Africa (1), Sweden (1), Taiwan (2), Turkey (3),* and *USA (7),* as depicted in Fig. Of these, multiple-site study data was collected in Germany, Canada, and Israel (Boehnke 2005), and Greece and Turkey (Dede 2013a, b). One study analyzed data from 60 nations and presented meta-level findings with country specific findings not described individually (Fang et al. 2016).

Fig. Frequency of publications

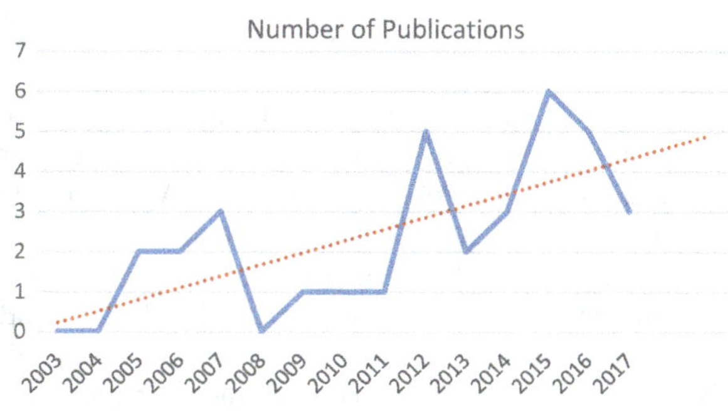

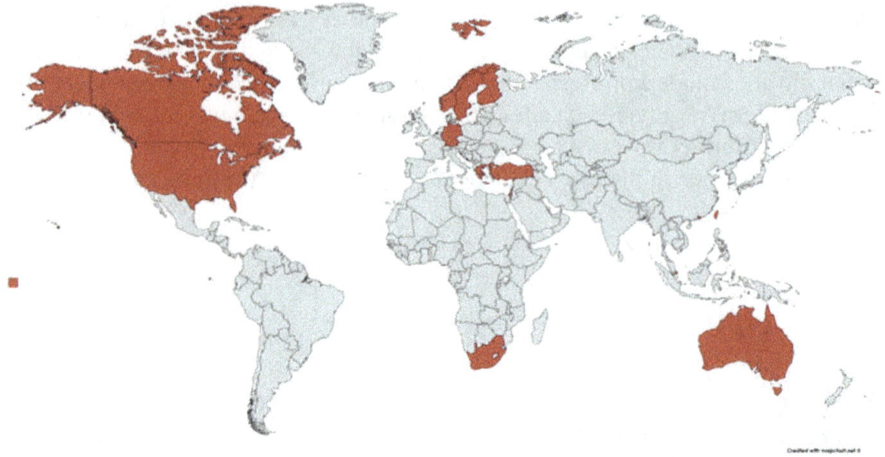

Fig. Map of the countries that studies have been conducted in

Which Stakeholders Are Represented in the Research?

While studies primarily collected data directly from the students, teacher data (Chouinard et al. 2007; Dede 2013a, b; Diemer et al. 2016; Federici and Skaalvic 2014; Haara and Smith 2012; Leu 2005; Metallidou and Vlachou 2010; Peng and Nyroos 2012), parent perspectives (Gniewosz and Noack 2012; Chouinard et al. 2007), and peers (Bissell-Havran and Loken 2009) were occasionally included. A total of 152,500 student participants were included in the 34 studies.

Student participants ranged from 7.5 to 18 years. The majority of studies focused on students who have reached adolescence rather than students in their early years. Two articles reported on a longitudinal studies that tracked students from Grade 6 until Grade 12 (Wang 2012), and from Grade 3 until Grade 12 (Simpkins et al. 2006).

What Is Known About the Development of Values?

Factors that may influence the development of values was a common theme amongst the studies. In particular, student-perceived ability was frequently examined, noted in five studies (Gniewosz and Watt 2017; Viljarants et al. 2016; Diemer et al. 2016; Gaspard et al. 2015; Metallidou and Vlachou 2010). One study reported on the development of student values in mathematics as a function of maternal and paternal values in mathematics (Gniewosz and Noack 2012), one study reported on the supportive role of peers and students perceptions of their peers valuing in mathematics (Bissell-Havran and Loken 2009), and one study reported on the influence of mathematics classroom experiences over the development of students values (Wang 2012). One study examined the profile of a resilient learner who has succeeded in mathematics despite adversity, and described a female learner in which the language of classroom instruction was not spoken at home, who places a high value on mathematics (Frempong et al. 2016).

The utility of modelling activities compared to traditional problems solving and the subsequent impact of the development of student values was reported in two studies (Dorak 2012; Haara and Smith 2012). The role of values in relation to mathematics anxiety was explored in one study (Henschel and Roick 2017), in relation to the prediction of motivation and effort in mathematics was explored in five studies (Andersen and Cross 2014; Berland and Steingut 2016; Federici and Skaalvik 2014; Hsiang 2017; Penk and Schipolowski 2015) and more specifically in relation to effort in mathematics homework in one study (Trautwein et al. 2006).

One study examined self-regulated strategy use in elementary mathematics and specifically considered the role of student valuing in this context (Chatzistamatious et al. 2015). The authors reported that enjoyment and positive valuing of the importance of mathematics as a school subject are necessary for mastery goals to have a positive effect on students' use of self-regulated strategies in mathematics. Elsewhere, the relationship between self-concept and utility values in the prediction of educational outcomes, including persistence in mathematics, was reported in three studies (Guo et al. 2015; Andersen and Ward 2014; Fries et al. 2007). One study reported that the relationship between achievement value and academic achievement performance was not overly strong and suggest that achievement values may play an ambiguous role in generating high academic performance (Boehnke 2005).

How Consistent Are the Findings Reported in the Studies?

Gender was explored in five studies, with two studies describing gender differences (Henschel and Roick 2017; Gaspard et al. 2015) and three describing consistency in findings for both genders (Muis et al. 2015; Guo et al. 2015; Simpkins et al. 2006). One study specifically described omitting special needs students from their data (Penk and Schipolowski 2015), one study noted that 30% of the students had an Individualised Education Plan (IEP) (Muis et al. 2015), and one study included one general education class and one special education class (Peng and Nyroos 2012). Difference in values held by general education students when compared to students in special needs education was described in one study (Peng and Nyroos 2012).

Conclusion and Implications

The aim of this study was to provide a mapping of the empirical research that has been conducted to date, and to use the research findings to inform the evolving definition of values in mathematics education. Findings from the data set as a whole suggest that the role of students' levels of motivation and effort are of prime importance to understanding student values in mathematics. Developing an understanding of how students perceptions of their mathematical abilities impact their valuing has been given almost equal attention in the research. The utility value of mathematics, and how this relates to student valuing has been frequently investigated. The most frequently represented countries in the data set are Germany and the USA.

Although the majority of studies included high school students, research conducted in lower grade levels has suggested students with higher cognitive ability and greater motivation hold high value beliefs in mathematics (Metallidou and Vlachou 2010), and that positive value beliefs are necessary for mastery goals to be effective (Chatzistamatiou et al. 2015).

Drawing upon the current findings, it appears that there is general agreement that suggests that values are a reflection of motivation and effort largely shaped by perceived ability. Additionally, the role of perceived utility appears central to the subsequent values held by the individual. When considered using the traditional theory of psychology, the classic partition of the mind is viewed in terms of three functions: cognition, emotion, and conation—the will or volitional component that drives an individual in his or her application to a given task. To date it appears that values researchers have explored cognition, and emotion largely operationalizing the investigations in terms of ability, achievement, self-concept, motivation and effort. Less is currently known about the role of conation in shaping values.

In contrast to frequent reports of the significance of positive valuing of high achievement in relation to favourable academic performance outcomes, the exploratory study conducted by Boehnke (2005) has highlighted a weak relationship. More specifically, Boehnke has explained that the impact of achievement values on performance is always indirect, elaborating that such valuing infact impact achievement related self-esteem. Boehnk demonstrated a two-field influence of achievement value on grades arguing both a positive and negative impact on achievement related self-esteem. While noting the line of research as exploratory, Boehnke has alerted educational researchers to the possibly ambiguous role achievement values play in the generation of high academic performance. Further, inconsistent findings are included in this data set regarding the role of gender differences in relation to values in mathematics.

To date little research investigating changes in values in mathematics teaching and learning has been conducted, either across time or as environmental changes attributed to migratory and immigration trends occurs. One large group study conducted in Taipei, Singapore, and America across both Grade 4 and Grade 8 reported that student values and competence beliefs decrease over time (Hsiang-Wei 2017). More specifically, we can understand this to mean that the less students like mathematics over time, the less they believe they are competent in this subject, and deduce that their positive valuing of mathematics as a subject decreases. Ongoing research conducted across time intervals, and in a variety of location appears highly warranted, to better identify patterns in changes in valuing and subsequently develop intervention strategies aimed to promote optimal outcomes in student mathematics achievement.

Immigration trends around the world has meant that many families are relocating to new countries. Research into whether existing values in mathematics education are retained, or new values developed is one important line of query. In particular, questions arise regarding whether opportunities in new environments are able to be accessed by newly arrived families, and do these new settings contribute to positive associations with studying mathematics.

The main data collection method utilized by these studies has been a self-report questionnaire instrument, of either student or other stakeholder measures. Few studies have included other data sources, such as classroom observations or interviews, and even fewer studies have included multiple stakeholder responses. Few longitudinal studies have been conducted (Muis et al. 2015; Simpkins et al. 2006). Further, there is a paucity of research that has adopted a pre- and post- assessment of valuing in conjunction with mathematics classroom teaching intervention, skill-building intervention, or home-work intervention. Future research that addresses these knowledge gaps appears highly warranted.

Countries around the world are encountering multi-culturism in new ways. With this comes new challenges to mathematics education, and arguably the field of values is increasingly pertinent to successful teaching and learning of mathematics. Many students face tremendous challenges before entering the mathematics classroom – language barriers, ethnic or racial tension, economic hardship to name but a few. In these instances, it is increasingly important to address a variety of factors in the broader environment that may impact experiences inside the mathematics classroom. Values would be one of these factors, given that these are often shaped externally in the societies and communities in which the students operate, but espoused in the classrooms as the students negotiate on a day-to-day basis the border crossings between home and school.

A significant limitation of this review is that only English language publications have been included. Many values researchers are active throughout Asia, and publications in various languages exist. A recent study by Peng and Nyroos (2012) published in Korean in The Mathematical Education journal of the Korean Society for Mathematics Education is one such example.

Appendix: Summary of Studies

Authors	Year	Sample size	Year level	Mean age	Gender	Setting	Location	Aim of study	Data collection	Findings
Gniewosz and Watt	2017	n = 398 n = 414 n = 459 N = 1,271	Grade 7–10	13.25 years 12.36 years 14.41 years	44.9% girls 43.6% girls 42.9% girls	Upper middle-class coed secondary school; 3 cohorts	Sydney, Australia	Student perceptions of parents and teachers' overestimation of ability	Mathematical task values: intrinsic and utility	Perceived encouragement conveyed by student-perceived mathematical ability beliefs of parents and teachers promote positive mathematics task values development
Henschel and Roick	2017	N = 368	Grade 4	9.4 years	52% girls	Elementary school	Germany	Value of learning and studying mathematics for its own sake	Intrinsic domain value—6 questions; Extrinsic achievement value—3 questions	1. Anxieties correlate stronger with control beliefs than domain values; 2. Achievement values not related with anxiety; 3. Girls reported lower intrinsic values; 4. No gender differences achievement outcomes

(continued)

(continued)

Authors	Year	Sample size	Year level	Mean age	Gender	Setting	Location	Aim of study	Data collection	Findings
Hsiang-Wei	2017	N = 44,667	Grade 4 and 8	10–14		TIMMS 2011 database	Taipei, Singapore, America	Effect of motivation and engagement on achievement	Motivational framework including intrinsic value	Students' values and competence beliefs decrease over time
Berland and Steingut	2016	N = 113	K–12	n/a	n/a	8 schools—Engineer Your Word 2012–2013	South Central USA	How student perceptions of value of math and expectancy for success relate to effort	Survey of expectancy, value and effort towards mathematics and science in engineering design challenges	Subjective task value was found to significantly predict effort towards mathematics Researchers argue the need for educators to help students to recognize the value of each domain within STEM environment
Diemer, Marchand, McKellar, and Malanchuk	2016	MADIS Wave 3 N = 618	End of Grade 8	n/a	45.6% girls	23 Public middle schools; African-American sample	Prince George County, Maryland, USA	Relevant instruction, self-concept of ability, task value, and achievement	Teachers' differential treatment: 1item of questionnaire measure youths' mathematics task value (expectancy-value framework)	Findings suggested that self-concepts of ability are predictive of task value and play a role in achievement over time

(continued)

(continued)

Authors	Year	Sample size	Year level	Mean age	Gender	Setting	Location	Aim of study	Data collection	Findings
Fang, Xu, Grant, Stronge and Ward	2016					TIMMS 2011 data	60 Counties	Relationship between national culture, creativity	World Values Survey (WVS)—traditional/secular and survival/self-expression dimension	Countries high on secular dimension performed higher on TIMMS student achievement Countries high on self-expression outperformed those high on survival
Frempong, Visser, Feza, Winnaar, and Nuamah	2016	N = 11,969	Grade 9			298 Schools; refined to public schools only	South Africa	Explored what drives the success of resilient learners	TIMMS 2011 public schools; explored gender, how often test language spoken at home, students' attitudes, bullying, parent education level, homework	Identified typical resilient learner: girl who does not speak classroom instruction language at home, tends to value and like mathematics and expressed confidence about their ability to learn mathematics
Viljaranta, Aunola, and Hirvonen	2016	N = 156	Grade 1	7.5 years	79 boys 77 girls	Three schools	Northern Finland	Children's intrinsic value, self-concept of ability, performance	Interview using Task-Value Scale for Children	Five groups: positive, negative, math-motivated, reading-motivated, low intrinsic value but high belief. Motivational patterns associated with students' level of performance at start and throughout grade 1

(continued)

(continued)

Authors	Year	Sample size	Year level	Mean age	Gender	Setting	Location	Aim of study	Data collection	Findings
Chatzistamatiou, Dermitzaki, Efklides, and Leondari	2015	N = 344	Grade 5 and Grade 6	11–12 years	163 girls 181 boys	Elementary school (medium socio-economic status)	Greece	Self-regulatory strategies, achievement goals, self-efficacy, value, enjoyment	Value: three items assessed students value beliefs about importance of mathematics	Results showed that students' positive self-efficacy, value beliefs, and enjoyment of mathematics are necessary for mastery goals to have a positive effect on mathematics strategy use
Gaspard, Dicke, Flunger, Brisson, Hafner, Nagengast, and Trautwein	2015	N = 1,916	Grade 9	14.62 years	53.5% girls	25 Academic-track Gymnasium schools, 82 classrooms	Baden-Wurttenberg, Germany	Relevance intervention in the classroom, assessment based on Expectancy-Value theory	37 items addressed all four components of EVT: Intrinsic value 4 items, Attainment value 10 items, Utility value 12 items, Cost value 11 items	Classroom intervention assessed via self-reports Compared to control comparison, classes in the quotation condition reported higher utility value, attainment value, and intrinsic value and classes in the text condition reported higher utility value

(continued)

(continued)

Authors	Year	Sample size	Year level	Mean age	Gender	Setting	Location	Aim of study	Data collection	Findings
Gaspard, Dicke, Flunger, Schreier, Hafner, Trautwein, and Nagengast	2015	N = 1,868	Grade 9	14.62 years	53.3% girls	25 Academic-track Gymnasium schools, 82 classes	Baden-Wurttenberg, Germany	Gender differences: Expectancy-Value Theory: attainment value, intrinsic value, utility value, cost	37 items addressed all four components of EVT	Conceptual differences of value beliefs, achievement, personal importance, utility, effort, emotional cost and opportunity cost. Mean differences favoured boys
Guo, Marsh, Parker, Morin, and Yeung	2015	TIMMS 1999 5,179 2003 4,972 2007 3,470	Grade 8	14.4 years	49.3% girls (1999) 50.4% girls (2003, 2007)	Stage 1: schools Stage 2: classroom selected from stage 1	Hong Kong	Expectancy-Value, gender, and socio-economic background as predictors of achievement	TIMSS scale of students' positive affect (intrinsic value) and TIMSS scale of students valuing (utility value); academic achievement, educational aspirations, background	Self-concept and lower utility values predictive of outcomes, boys' and girls' similar levels of self-concept and values, girls' higher achievement and educational aspirations, socioeconomic status linked to aspirations for boys

(continued)

(continued)

Authors	Year	Sample size	Year level	Mean age	Gender	Setting	Location	Aim of study	Data collection	Findings
Muis, Psaradellis, Lajoie, Di Leo, and Chevrier	2015	N = 79 (30% students had IEP)	Grade 5	11 years	34 girls 45 boys	Eclectic mix of low—high income families	Canada	Prior knowledge, emotions, task values, academic control, activity	Task value measured at four points in time, used to measure students value in learning mathematics in general as well for problem solving	No gender differences found for task value. Perceived control and value served as important antecedents to the epistemic and activity emotions students experience during problem solving
Penk and Schipolowski	2015	N = 42,298	Grade 9	15.6 years	49.8% girls	Nationally representative sample excluding special ed students	Germany	Test taking motivation questionnaire (pre-post achievement test)	Values measured: Importance, Interest, Anxiety	Value component more important than expectancy component for prediction of effort. When viewed together value and effort taken explained over a quarter of the variance in mathematics scores

(continued)

(continued)

Authors	Year	Sample size	Year level	Mean age	Gender	Setting	Location	Aim of study	Data collection	Findings
Andersen and Cross	2014	N = 19,259 after missing data omitted	Grade 9	n/a	n/a	944 public and private schools, 27 students per school	10 states across USA	Explored whether high ability students are more motivated than other students	National Center for Education Statistics Expectancy-value theory: achievement, self-efficacy, attainment value, utility value, interest-enjoyment	Identified high self-efficacy with lower utility value and high utility value with lower self-efficacy. 41% of high-ability students had high motivation, 15% of high-ability students had low motivation
Andersen and Ward	2014	n = 221 Black n = 351 Hispanic n = 1,185 White N = 1,757	Grade 9	n/a	123 girls; 180 girls; 546 girls	944 public and private schools, 27 high ability students per school;	10 states across USA	Comparison of STEM persistence plans of high-ability Black, White, and Hispanic students	Expectancy-value scale: self-efficacy, identity; attainment scale: utility value 4 items, intrinsic value 4 items, cost value 4 items in mathematics and science	Black: persisters significantly higher than non-persisters in achievement value. Hispanic: persisters significantly higher than non-persisters in STEM utility value; White: persisters scored significantly higher than non-persisters in mathematics attainment value

(continued)

(continued)

Authors	Year	Sample size	Year level	Mean age	Gender	Setting	Location	Aim of study	Data collection	Findings
Federici and Skaalvik	2014	N = 309	Grade 9 and 10	n/a	51.8% girls 48.2% boys	Two middle schools 3 classes Grade 9 5 classes Grade 10	Large city in Norway	Subjective task values in relation to students' perceptions of teacher support and student effort	Norwegian language instrument: teacher instrumental support—7 items, utility value—5 items, cost value—3 items, intrinsic value—6 items, effort—3 items	Instrumental support directly positively related to utility and intrinsic value but only indirectly related to perceived cost value of mathematics. Where instrumental support was indirectly related to effort the relation was mediated by students' perceptions of task values
Dede	2013 (a)	N = 22	Primary and Secondary	n/a	n/a	13 German teachers; 9 Turkish teachers	Northern Germany; Central Anatolia, Turkey	Explored underlying values of Turkish and German mathematics teachers	Values in Mathematics Teaching in Turkey and Germany (VMTG): semi-structured interviews and field notes	Identified four values categories: Productivity, socialization, flexibility/authority, gender differences. Gender differences more important to German teachers; Turkish teachers attached great importance to student productivity

(continued)

(continued)

Authors	Year	Sample size	Year level	Mean age	Gender	Setting	Location	Aim of study	Data collection	Findings
Dede	2013 (b)	N = 60	Primary and Secondary		24 Female 26 Male	27 German teachers; 33 Turkish teachers	Berlin (27) Sivas (31) Ankara (2)	Does mathematics teachers experience and nationality influence values	Five point likert type questionnaire	Teaching experience between the countries has a significant effect on the values, and both Turkish and German mathematics teachers' level of experience does not have a significant impact on their values
Doruk	2012	n = 34 grade 6 n = 24 grade 7 N = 58	Grade 6 and 7	n/a	28 boys 30 girls	Primary school low socio-economic status	Ankara, Turkey	Processes that teach mathematics and educational values	Researcher observations, student worksheets, videos, reports and semi-structured interviews; thematic analysis: general education values; mathematics values	Student development of models, defending functionality and discussion effective for developing responsibility values. Integration of teamwork and communication develop social and cultural values

(continued)

(continued)

Authors	Year	Sample size	Year level	Mean age	Gender	Setting	Location	Aim of study	Data collection	Findings
Gniewosz and Noack	2012	n = 874 mothers n = 733 fathers	Grade 5 and 6	10.6 years and 12.1 years	49.1% male, 47.4% male	15 regelschule and 15 gymnasium schools; 60 classrooms	Thuringen, Germany	Intergenerational transmission of the valuing of mathematics within family	Student and parent questionnaire at two time points: beginning Grade 5 and mid Grade 6; measured attainment value, intrinsic value, utility value	Group 1 mothers valuing predicted students' own mathematics values; Group 2 fathers valuing predicted students' own valuing of mathematics. Dyad gender predicted group membership
Haara and Smith	2012	N = 2	Grade 4 and Grade 8 teachers	9—10 years and 13—14 years	n/a	Upper primary school and Lower secondary school	Norway	In-service course to introduce a values-based approach to teaching	interviews, videos of observation and teachers' reactions to their videos, logs, open-ended questionnaire	VaKE provided opportunity for increased use of practical teaching activities, but also showed how good intentions of changing practice may be restrained by beliefs and prior experiences

(continued)

(continued)

Authors	Year	Sample size	Year level	Mean age	Gender	Setting	Location	Aim of study	Data collection	Findings
Peng, and Nyroos	2012	N = 2	Grade 7 and 8	n/a	n/a	Prestigious city school; general ed and special ed classrooms	Northern Sweden	What is valued by students in each group, what is valued by teachers in each group	Lesson observations, student focus group interviews, and teacher interviews	General Ed students three most cited values: explanation, quietness, and personalised help; Special Ed students three most cited values: independence, relaxation, and explanation
Wang	2012	N = 14,236	Data collected at five time points	Grade 7: 12.4 years	54% female (G7)	124 mathematics classrooms drawn from 12 public schools	Southeastern Michigan, USA	Predictors of student choices to enroll in highschool, and mathematical occupational aspirations	Grades, number of courses, aspirations, motivational beliefs: expectancies—5 items, subjective task values i. importance—3 items ii. Interest—3 items,	Mathematics classroom experiences predicted expectancies and values, which in turn predicted the number of high-school mathematics courses taken and students career aspirations in mathematics related areas

(continued)

(continued)

Authors	Year	Sample size	Year level	Mean age	Gender	Setting	Location	Aim of study	Data collection	Findings
Yazici	2011	N = 359	Pre-service teachers	Primary and Secondary			Turkey	Pre-service teachers' mathematical values and their teaching anxieties	Mathematics Teaching Anxiety Scale—23 items; Mathematics Values Scale—34 items (Positivist values and Constructivist values)	Constructivist value preferences of pre-service teachers directly affect their mathematics teaching anxieties; no significant relationship between positivist values and mathematics teaching anxiety
Metallidou and Vlachou	2010	n = 114 Grade 5 n = 149 Grade 6	Grade 5 and Grade 6	10 years 7 months and 11 years 9 months	133 girls 130 boys	13 public primary schools,	Greece	Explored the role of task-value beliefs in childrens' self-regulated learning	Task-value beliefs—9 items, self-efficacy, test anxiety. Teacher ratings of student achievement, meta-cognitive knowledge, student self-regulation	Students with high value beliefs in mathematics were described as more cognitively, metacognitively, and motivationally competent learners as compared to students with lower value beliefs

(continued)

(continued)

Authors	Year	Sample size	Year level	Mean age	Gender	Setting	Location	Aim of study	Data collection	Findings
Bissell-Havran and Loken	2009	n = 207	Grade 8	13 years	114 boys 113 girls	Rural school, 21% economically disadvantaged	Mid-Atlantic, USA	Explored the role of friendships in academic self-competence and intrinsic values in Mathematics	Measures: friends support, intrinsic values for English and Mathematics—4 items (each), academic self-competence	Analyses predicting intrinsic value for mathematics and English provided weaker evidence of an interaction. Students also perceived that their friends valued academics significantly less than the friends actually reported
Chouinard, Karsenti, and Roy	2007	N = 759	Grade 7 – 11	12–18 years	389 boys 370 girls	4 public highschools	Montreal, Canada (French speaking)	Relationship between the beliefs, utility value and achievement goals	Confidence—6 items, utility—5 items, mastery—8 statements, performance—6 statements, work-avoidance—6 statements; effort—3 items	Mastery goals had significant impact on student's effort in learning mathematics. Competence beliefs, utility value, achievement goals and effort not significantly influenced by age and gender

(continued)

(continued)

Authors	Year	Sample size	Year level	Mean age	Gender	Setting	Location	Aim of study	Data collection	Findings
Frenzel, Pekrun, and Goetz	2007	n = 2,053	Grade 5	11.7 years	1,036 boys 1,017 girls	42 schools mix of socio-economic and cultural background	Bavaria, Germany	Gender difference in emotion towards mathematics	Measures: competence belief, value belief, mathematics emotions and mathematics grades	Girls reported significantly less enjoyment and pride, more anxiety, hopelessness and shame. Female emotional pattern due to the girls' low competence beliefs and domain value and high achievement value
Fries, Schmid, and Hofer	2007	N = 704	Grade 6 and 8	13.5 years	48.4% boys 51.4% girls	9 schools from all 3 school types 29.5% immigrant	Ludwigshafen, Germany	Relationship between value orientations, valences, and academic achievement	Similarity to students prototypes for achievement value and well-being value; Intrinsic incentives—4 items, extrinsic incentives 4 items; Actual grades	Results showed that school grades were significantly predicted by value orientation
Simpkins, Davis-Kean, and Eccles	2006	N = 227	Longitudinal data Grade 5, 6, 10, 12	8.33 years Grade 3	54% girls	12 public schools from three school districts	Michigan, USA	Mathematics/science choices and beliefs from child-hood—adolescence	Expectancy-value at Grade 6 and 10	Youths' mathematics and science activity participation predicted their expectancies and values which in turn predicted the number of high school courses taken No gender differences

(continued)

(continued)

Authors	Year	Sample size	Year level	Mean age	Gender	Setting	Location	Aim of study	Data collection	Findings
Trautwein, Ludtke, Kastens, and Koller	2006	Study 1 N = 2,712	Grade 5, 7, 9	13.37 years	49.5% girls	158 classes representing 11 schools	Large city in Germany	Age-related differences in students mathematics homework	Measures: homework effort—4 items; expectancy-value using self-concept of ability and intrinsic value—5 items; conscientiousness	Lower homework effort in higher grades. Intrinsic value on homework effort were higher in the older cohorts, whereas effects of the expectancy component were lower
		Study 2 N = 571	Grade 8 and 9	14.72 years	51.5% girls	44 classes from 10 schools	Large city in Germany	Power of motivation to predict effort	Effort—6 items; concentration—3 items; motivation—3 items, value—3 items; control—3 items, adaptability—3 items	Means for effort and value were lower for homework than for classwork with these differences being partly moderated by the students' conscientiousness

(continued)

(continued)

Authors	Year	Sample size	Year level	Mean age	Gender	Setting	Location	Aim of study	Data collection	Findings
Boehnke	2005	Germany n = 641 Canada n = 605 Israel n = 419 N = 1,665	Grade level not specified	14 years	Germany 336 girls Canada 301 girls Israel 205 girls	Germany 14 schools; Canada 4 schools; Israel 2 schools	Chemnitz, East Germany; Calgary, Canada; and Beer-Sheva, Israel	Explored the role of achievement value in mathematical achievement as measured by school grades and test scores	Measures:—most recent mathematics report grades, independent assessment using TIMSS; self-esteem—9 items; manifest anxiety—6 items; parental achievement expectancies—3 items; achievement value preferences—4 items	Relationship between achievement value and academic achievement performance as a behaviour measure is not overly strong. Findings proposed as a means to alert educational researchers to the possibly ambiguous role of achievement values in generation of high academic performance
Leu	2005	N = 1	Elementary mathematics	n/a	17 girls, 19 boys; 1 teacher	Teachers college affiliated Laboratory School	Taipei, Taiwan	Teachers pedagogical values and her students' perceptions	Weekly lesson observation; Observation of complete unit of instruction (geometry); occasional and intermittent observations	Goal of education; value of dealing with people and life; value of mathematics learning; value of mathematics teaching; value of mathematics education

References

*References marked with an asterisk indicate studies identified in the systematic literature review.

*Andersen, L., & Cross, T. L. (2014). Are students with high ability in math more motivated in math and science than other students? *Roeper Review: A Journal on Gifted Education, 36*(4), 221–234.

*Andersen, L., & Ward, T. J. (2014). Expectancy-value models for the STEM persistence plans of ninth-grade, high-ability students: A comparison between Black, Hispanic, and White students. *Science Education, 98*(2), 216–242.

*Berland, L. K., & Steingut, R. (2016). Explaining variation in student efforts towards using math and science knowledge in engineering contexts. *International Journal of Science Education, 38*(18), 2742–2761. https://doi.org/10.1080/09500693.2016.1260179.

Bishop, A. (1988a). Mathematics education in its cultural context. *Educational Studies in Mathematics, 19*(2), 179–191.

Bishop, A. (1988b). *Mathematical encultration: A cultural perspective on mathematics education.* Dordrecht, Netherlands: Kluwer.

Bishop, A. J., Seah, W. T., & Chin, C. (2003). Values in mathematics teaching: The hidden persuaders? In A. J. Bishop, K. Clements, C. Keitel, J. Kilpatrick, & F. Leong (Eds.), *International handbook of mathematics education* (2nd ed., pp. 715–763). Dordrecht, The Netherlands: Kluwer Academic Publishers.

*Bissell-Havran, J. M., & Loken, E. (2009). The role of friends in early adolescents' academic self-competence and intrinsic value for math and English. *Journal of Youth and Adolescence, 38*(1), 41–50.

*Boehnke, K. (2005). Value orientations in relation to mathematical self-esteem: An exploratory study of their role in mathematical achievement among German, Israeli, and Canadian 14-year-olds. *European Journal of Psychology of Education, 20*(3), 227–241.

*Chatzistamatiou, M., Dermitzaki, I., Efklides, A., & Leondari, A. (2015). Motivational and affective determinants of self-regulatory strategy use in elementary school mathematics. *Educational Psychology, 35*(7), 835–850.

*Chouinard, R., Karsenti, T., & Roy, N. (2007). Relations among competence beliefs, utility value, achievement goals, and effort in mathematics. *British Journal of Educational Psychology, 77*(3), 501–517.

Dede, Y. (2006). Values in Turkish middle school mathematics textbooks. *Quality & Quantity, 40,* 331–359.

*Dede, Y. (2013a). Examining the underlying values of Turkish and German mathematics teachers decision making processes in group studies. *Educational Science Theory & Practice, 13*(1), 690–706.

*Dede, Y. (2013b). The effect of German and Turkish mathematics teachers' teaching experience on mathematics education values: A cross-comparative study. *International Journal of Mathematics Education in Science and Technology, 44*(2), 232–252. https://doi.org/10.1080/0020739x.2012.714489.

*Diemer, M. A., Marchand, A. D., McKellar, S. E., & Malanchuk, O. (2016). Promotive and corrosive factors in African American students' math beliefs and achievement. *Journal of Youth and Adolescence, 45*(6), 1208–1225.

*Doruk, B. K. (2012). Mathematical modeling activities as a useful tool for values education. *Kuram ve Uygulamada Egitim Bilimleri. 12*(2, Suppl), 1667–1672.

Eklof, H. (2007). Self-concept and valuing of mathematics in TIMSS 2003: Scale structure and relation to performance in a Swedish setting. *Scandinavian Journal of Educational Research, 51*(3), 297–313.

*Fang, Z., Xu, X., Grant, L. W., Stronge, J. H., & Ward, T. J. (2016). National culture, creativity, and productivity: What's the relationship with student achievement? *Creativity Research Journal, 28*(4), 395–406. https://doi.org/10.1080/10400419.2016.1229976.

*Federici, R. A., & Skaalvik, E. M. (2014). Students' perception of instrumental support and effort in mathematics: The mediating role of subjective task values. *Social Psychology of Education, 17*(3), 527–540.

*Frempong, G., Visser, M., Feza, N., Winnar, L., & Nuamah, S. (2016). Resilient learners in schools serving poor communities. *Electronic Journal of Research in Educational Psychology, 14*(2), 352–367. https://doi.org/10.14204/ejrep.39.15038.

*Frenzel, A. C., Pekrun, R., & Goetz, T. (2007). Girls and mathematics–A "hopeless" issue? A control-value approach to gender differences in emotions towards mathematics. *European Journal of Psychology of Education, 22*(4), 497–514.

*Fries, S., Schmid, S., & Hofer, M. (2007). On the relationship between value orientation, valences, and academic achievement. *European Journal of Psychology of Education, 22*(2), 201–216.

Furuto, L. (2014). Pacific ethnomathematics: Pedagogy and practices in mathematics education. *Teaching Mathematics and its Applications, 33*(2), 110–121.

*Gaspard, H., Dicke, A., Flunger, B., Brisson, B. M., Hafner, I., Nagengast, B., et al. (2015a). Fostering adolescents' value beliefs for mathematics with a relevance intervention in the classroom. *Developmental Psychology, 51*(9), 1226–1240.

*Gaspard, H., Dicke, A., Flunger, B., Schreier, B., Hafner, I., Trautwein, U., et al. (2015b). More value through greater differentiation: Gender differences in value beliefs about math. *Journal of Educational Psychology, 107*(3), 663–677.

*Gniewosz, B., & Watt, H. M. (2017). Adolescent-perceived parent and teacher overestimation of mathematics ability: Developmental implications for students' mathematics task values. *Developmental Psychology, 53*(7), 1371–1383.

*Gniewosz, B., & Noack, P. (2012). Mamakind or papakind? [Mom's child or Dad's child]: Parent-specific patterns in early adolescents' intergenerational academic value transmission. *Learning and Individual Differences, 22*(4), 544–548.

*Guo, J., Marsh, H. W., Parker, P. D., Morin, A. J. S., & Yeung, A. S. (2015). Expectancy-value in mathematics, gender and socioeconomic background as predictors of achievement and aspirations: A multi-cohort study. *Learning and Individual Differences, 37,* 161–168.

*Haara, F. O., & Smith, K. (2012). Increasing the use of practical activities through changed practice. *The Mathematics Enthusiast, 9*(1–2), 77–110.

*Henschel, S., & Roick, T. (2017). Relationships of mathematics performance, control and value beliefs with cognitive and affective math anxiety. *Learning and Individual Differences, 55,* 97–107.

*Hsiang-Wei, K. (2017). The effects of motivational constructs and engagement on mathematics achievements; A comparative study using TIMSS 2011 data of Chinese Taipei, Singapore and the USA. *Asia Pacific Journal of Education, 37*(2), 135–149. https://doi.org/10.1080/02188791.2016.121682.

Kluckhohn, F. R., & Strodbeck, F. L. (1961). *Variations in value orientations*. Westport: CT, Greenwood Press.

Leu, Y. C. (2005). The encactment and perceptions of mathematics pedagogical values in an elementary classroom: Buddhism, Confucianism, and curriculum reform. *International Journal of Science and Mathematics Education, 3,* 175–212.

*Metallidou, P., & Vlachou, A. (2010). Children's self-regulated learning profile in language and mathematics: The role of task value beliefs. *Psychology in the Schools, 47*(8), 776–788.

*Muis, K. R., Psaradellis, C., Lajoie, S. P., Di Leo, I., & Chevrier, M. (2015). The role of epistemic emotions in mathematics problem solving. *Contemporary Educational Psychology, 42,* 172–185.

Parsons, T., & Shils, E. A. (1951). *Toward a generall theory of action*. Cambridge: MA, Harvard University Press.

*Peng, A., & Nyroos, M. (2012). Values in effective mathematics lessons in Sweden: What do they tell us? *The Mathematics Enthusiast, 9*(3), 409–430.

*Penk, C., & Schipolowski, S. (2015). Is it all about value? Bringing back the expectancy component to the assessment of test-taking motivation. *Learning and Individual Differences, 42,* 27–35.

Seah, W. T. (2018). Improving mathematics pedagogy through student/teacher valuing: Lessons from five continents. In G. Kaiser, H. Forgasz, M. Graven, A. Kuzniak, E. Simmt, & B. Xu (Eds.), *Invited Lectures from the 13th International Congress on Mathematical Education* (pp. 561–580). Switzerland, Springer: Cham.

Seah, W. T., & Andersson, (2015). Valuing diversity in mathematics pedagogy through the volitional nature and alignment of values. In A. Bishop, H. Tan, & T. Barkatas (Eds). *Diversity in mathematics education: Towards inclusive practices* (pp. 167–183). Switzerland: Springer

Seah, W. T., & Bishop, A. (2001) Teaching more than numeracy: The socialisation experience of a migrant teacher. In 1. Bobis, B. Perry & M, Mitchelmore (Eds.), *Numeracy and beyond: Proceedingsof the Twenty-Fourth Annual Conference of the Mathematics Education Research Group of Australasia* (Vol. 2, pp. 442–450). Turramurra, Australia: Mathematics Education Research Group of Australasia.

*Simpkins, S. D., Davis-Kean, P. E., & Eccles, J. S. (2006). Math and science motivation: A longitudinal examination of the links between choices and beliefs. *Developmental Psychology, 42*(1), 70–83.

Timms, M., Moyle, K., Weldon, P., & Mitchell, P. (2018). *Challenges in STEM learning in Australian schools*. Australian Council for Education Research: Literature and Policy Review.

*Trautwein, U., Lüdtke, O., Kastens, C., & Köller, O. (2006). Effort on homework in grades 5–9: Development, motivational antecedents, and the association with effort on classwork. *Child Development, 77*(4), 1094–1111.

*Viljaranta, J., Aunola, K., & Hirvonen, R. (2016). Motivation and academic performance among first-graders: A person-oriented approach. *Learning and Individual Differences., 49,* 366–372.

*Wang, M. (2012). Educational and career interests in math: A longitudinal examination of the links between classroom environment, motivational beliefs, and interests. *Developmental Psychology, 48*(6), 1643–1657.

*Yazici, E., Peker, M., Ertekin, E., & Dilmac, B. (2011). Is there a relationship between pre-service teachers' mathematical values and their teaching anxieties in mathematics? *Electronic Journal of Research in Educational Psychology, 9*(1), 263–282.

2

What Do Pāsifika Students in New Zealand Value Most for their Mathematics Learning?

Julia Hill, Jodie Hunter and Roberta Hunter

Abstract Achieving equity for all mathematical learners is an urgent challenge for educators. Within New Zealand, Pāsifika students are at a much greater risk of under-achievement than students from other ethnic groups (Caygill et al. in TIMSS 2015: New Zealand Year 5 Maths results. Comparative Education Research Unit, Ministry of Education, Wellington, 2016a; Caygill et al. in TIMSS 2015: New Zealand Year 9 Maths results. Comparative Education Research Unit, Ministry of Education, Wellington, 2016b). In this chapter, we examine and explore the mathematics educational values of middle school Pāsifika students in New Zealand based on their significance in contributing to more effective and equitable mathematics learning. Drawing on survey responses and individual interviews with 131 Year Seven and Eight Pāsifika students, we highlight the most frequently espoused mathematics educational values as utility, peer collaboration/group-work, effort/practice, and family/familial support. Results from this study provide insight into what is valued by Pāsifika students and the types of classroom culture and pedagogy which could be developed to align with these students' values. The wider implications of the study address the need for educators to examine the mathematics educational values of minority students in order to provide equitable mathematics classrooms.

Keywords Mathematics educational values · Equity · Middle years · Pāsifika

J. Hill · J. Hunter (✉) · R. Hunter
Institute of Education, Massey University Albany, Private Bag 109204, Auckland, New Zealand
e-mail: j.hunter1@massey.ac.nz

J. Hill
e-mail: juliaLhill@gmail.com

R. Hunter
e-mail: r.hunter@massey.ac.nz

Introduction

Both in New Zealand and internationally, achieving equity for all mathematical learners is an urgent challenge for educators. New Zealand classrooms, like those in many other countries are becoming increasingly diverse. Pāsifika peoples are a fast growing population with the proportion of Pāsifika[1] students in New Zealand schools doubling over the past decade (Ministry of Education (MoE) 2006a). Pāsifika describes a multi-ethnic group of indigenous peoples from Pacific Island nations, including both those who were born in New Zealand and those who have migrated from the Pacific Islands. This group identifies themselves with the islands and/or cultures of the Cook Islands, Samoa, Tonga, Tokelau, Niue, Fiji, Tuvalu, and the Solomon Islands (Coxon et al. 2002).

Currently, New Zealand has one of the largest mathematical achievement gaps related to ethnicity across developed countries, with Pāsifika students in New Zealand at a much greater risk of underachievement (Education Assessment Research Unit and New Zealand Council for Educational Research 2015; Caygill et al. 2016a, b). Disruption of these trends are required if equitable outcomes are to be achieved. We argue that culturally responsive mathematics experiences, that is, recognition and use of students' cultural capital in all aspects of teaching and learning (e.g., see Gay 2010; Ladson-Billings 1994) has the potential to enhance equitable outcomes for Pāsifika students. A body of research studies (e.g., Civil and R. Hunter 2015; J. Hunter et al. in press) show improved equitable educational outcomes can be achieved when we attend to Pāsifika culture and values in the classroom. However, there appear to be limited research studies which specifically explore the mathematics educational values of Pāsifika students in New Zealand mathematics classrooms.

We define values as the fundamental "convictions which an individual has internalised as being the things of importance and worth" (Seah and Andersson 2015, p. 169) which act as "general guides to behaviour or as points of reference in decision-making or the evaluation of beliefs or actions" (Halstead 1996, p. 5). Mathematics educational values relate specifically to mathematics learning and pedagogy. They take place in the context of activities and decisions that are made to enhance the learning and teaching of mathematics (Bishop 1996). These values can influence student preference for types of learning activities and pedagogy used within the classroom. They are also highly sensitive to cultural influences and can vary depending on the culture of the learner (Barkatsas and Seah 2015). The authors of this chapter chose to focus on mathematics educational values based on the contribution of these values to more effective and equitable mathematics learning for Pāsifika learners, a minority group within New Zealand. Specifically, we explore the following research questions, firstly: What are the most important mathematics educational values espoused by Pāsifika learners in New Zealand classrooms? Secondly, what are the wider implications of investigating the mathematics educational values of minority students for countries with diverse cultural groups within their educational systems?

---

[1] 'Pāsifika' are a multi-ethnic group of indigenous peoples from the Pacific Islands, see page 2 and 3.

Pāsifika Peoples and Valuing

The Pāsifika Education Plan (MoE 2013) is a policy document that was developed by the New Zealand Ministry of Education to highlight the strategic direction planned to improve Pāsifika education outcomes in New Zealand. It covers both the compulsory education sectors as well as early learning through to tertiary education. Within this document, the key Pāsifika cultural values are identified as respect, reciprocity, service, inclusion, spirituality, leadership, love, belonging, and family. Although there have been a number of similar policy documents over the past decade (MoE 2006b, 2009), there appears to be limited values research related to Pāsifika peoples, particularly from the perspective of the learner or self-reporting of values. While there are no studies specifically exploring the types of mathematics education values held by Pāsifika learners in New Zealand, there are several studies which provide some insight into valuing by this group of learners.

Anthony (2013) explored students' perspectives about what it meant to be a "good" mathematics teacher and student. Anthony reported that Pāsifika students valued a "good" teacher as someone who cared about his/her students, and who provided clear explanations. In terms of perceptions about a "good" student, Anthony found that these students endorsed a greater proportion of collaborative values (e.g., sharing, mathematical community, respect) than the other ethnic groups in the study. Likewise, other studies (Averill 2012; Hāwera et al. 2007; Hunter and Anthony 2011; Sharma et al. 2011) investigated Pāsifika students' perspectives on learning experiences in the mathematics classroom. These researchers affirmed that Pāsifika mathematics students often endorsed values that were reflective of their collectivist cultural values. For example, their responses indicated that the students valued respect and positive relationships with peers/teachers, reciprocity and helping others, collaboration—group-work, and family support as important for their mathematics learning.

Methodology

The data used within this chapter are drawn from a larger study (Hill 2017) involving a mixed methods approach. A focus was placed on the use of student voice to better understand what students identifying with four different ethnic groups (East Asian, European, Māori, and Pāsifika) in New Zealand valued as important for their mathematics learning.

The participants in the larger study were 227 middle school (Years 7 and 8) students from four state schools. However, the focus in this chapter is on 131 Pāsifika students from three schools. All of these Pāsifika students attended low socio-economic, high poverty, urban schools. These three schools had been involved in an ongoing professional development and research project entitled *Developing Mathematical Inquiry Communities* (DMIC) (R. Hunter et al. in press). The DMIC project is

Table: The twelve mathematical activities/statements and their value indicators

Mathematical learning activity or statement	Value indicator
I learn more in maths by working with other children	Peer collaboration/group-work
Maths involves looking for different ways to find the answer	Flexibility with strategies
It is important to respect and like my maths teacher and for them to respect me	Respect
It is important for maths to be useful in real life or for my future	Utility
To be good at maths I need to practice with lots of questions	Effort and practice
I cannot be good at maths without the support/love/guidance of my family	Family/familial support
Maths needs to be clear and make sense to me	Clarity and understanding
It is important to talk about my ideas in a group or with my partner	Peer collaboration/communication
If I can't solve a difficult maths problem I need to keep working at it	Persistence
It is important to feel like I belong in my maths class	Belonging
It is important to get the correct/right answer in maths	Accuracy
My maths teacher needs to explain it to me properly so that I understand	Teacher explanations/clarity

implemented in schools that serve the most disadvantaged communities in New Zealand and is a transformative re-invention of pedagogical practices designed to support teachers' development of culturally responsive teaching (Gay 2010) and ambitious mathematics pedagogy (Kazemi et al. 2009).

Middle school students were the focus of this study because it is during this critical period that many students experience a negative and detrimental affective shift, with a decline in their academic engagement (Attard 2011; Grootenboer and Marshman 2015). The students were asked to individually complete a survey where they ranked 12 mathematics educational values from most important to least important, by numbering each value from one through to twelve. All the values used in the survey were derived from research literature and policy documents (e.g., Clarkson et al. 2000; M Œ 2013; Seah and Wong 2012). As children may find it difficult to relate and respond directly to particular values, each value was incorporated into a specific mathematical learning activity or statement (see Table).

For example, the statement "to be good at maths I need to practice with lots of questions" was understood to indicate the value of effort and practice in mathematics. During an individual follow-up interview, all students were asked to provide reasons for why they had selected the three most important values.

For the larger study, results from the survey were exported into Statistical Package for Social Scientists (SPSS) in order to investigate the statistical relationships among variables. For the findings reported in this chapter, we analysed how many students ranked each of the twelve values in their top three values to determine the degree of importance of each mathematical education value. As this is the first dedicated study into this area, more sophisticated analysis will be reported in later publications. All interview data was wholly transcribed and analysed through Nvivo software guided by a grounded theory approach in which codes, categories, patterns, and themes were developed. For example, a student response explaining why the value of collaboration/group was ranked highly was: "Because you can help other children" which was coded into the node of helping others. Another response: "They [peers] help me more, like get more confident" was coded into the node of confidence.

Findings and Discussion

Within this section to determine what P¯asifika students' value most in relation to mathematics educational values, we will identify the four highest ranked (that is, the most important) mathematics educational values (see Table) and explore student explanations of why they choose these values.

Utility

The statement "*It is important for maths to be useful in real life or my future*" was used as a value indicator for the mathematics educational value of utility. This value had the greatest proportion of students ranking it within their first three values with 47% ($n = 61/131$) of the students ranking it as one of their most important values.

More than half of the students ranking this value in their top three ($n = 33/61$) viewed the value of utility as important for their future education or career goals. This ranged from students equating engaging in mathematics as necessary for employment: *You need to be good at maths in order to get a job,* to students identifying specific occupations (e.g., accountant, police officer, banker, teacher) and the role of mathematics within this career choice. Other students identified the utility of mathematics as necessary for future study: *You can learn lots of maths for your future*

Table: The highest ranked mathematics educational values amongst P¯asifika students

Value	Percentage of Pāsifika students who ranked the value in their top three (n)
Utility	47% (61/131)
Peer collaboration/group-work	37% (48/131)
Effort and practice	24% (31/131)
Family/familial support	24% (31/131)

so that you might go to university. Alternatively, students ($n = 20/61$) identified the value of utility as important for everyday activities (e.g., cooking, shopping, building, money). For example, one student stated: *My Mum said if I never learn maths properly I won't know how to pay bills or count money*. Another student explained: *Every time I go with my Mum to shopping, we always need to figure out the discounts*. Interestingly, these statements suggest that utilitarian values can be influenced by the values held by parents and caregivers and also highlight the impact of out of school experiences.

Valuing utility means that these students desired mathematics which was either practical, or relevant to their own lives or the world around them. Students desired mathematics which had a purpose, that related to everyday activities, or which impacted upon their future success. Given the growing technological and digital economy (Ministry of Business Innovation and Employment 2017), there is a strong message both within New Zealand and internationally that mathematics is important for future employment and economic advancement (Gravemeijer et al. 2017). A previous study by Young-Loveridge et al. (2006) found that an overwhelming majority of students held beliefs about the importance of mathematics for the future, suggesting that utilitarian mathematics values are reflective of societal values and not necessarily distinguishable by cultural differences.

International values research studies (e.g., Barkatas and Seah 2015; Österling and Andersson 2013) also indicates the dominance of mathematical utilitarian values. For example, Österling and Andersson reported that Swedish middle school students highly valued "connecting mathematics to real life" (p. 22). Similarly, Barkatsas and Seah found the favourite mathematical tasks reported by students across Australia and several Chinese states involved real life scenarios. The strong utility values held by students both in the current study and other published research (e.g., Barkatas and Seah 2015; Österling and Andersson 2013) reaffirms the need for mathematics teachers to provide authentic learning experiences with opportunities for students from all cultural groups to apply concepts and skills to real life scenarios.

Peer Collaboration/Group-Work

To investigate the mathematics educational value of peer collaboration/group-work, the statement *"I learn more in maths by working with other children"* was used. Thirty-seven percent ($n = 48/131$) of students ranked this statement within their top three values.

In the follow-up interview, it appeared that there was a link between this mathematics education value and a key Pāsifika cultural value of reciprocity. Students viewed collaboration and group-work as providing reciprocal learning opportunities: *Because we learn more with each other.* Many students ($n = 34/48$) referred to the benefits of gaining new knowledge and learning new strategies from their peers as well as having their own errors highlighted. For example, one student explained: *So you can get different strategies*, another stated: *Because if I get something wrong, they will correct me.* Conversely, a smaller group of students ($n = 8/48$) also spoke of the opportunity they had themselves to help others. For example, one student explained: *So when they are stuck I can help them.* These types of explanations can also be linked to the cultural value of service as identified in the Pāsifika Education Plan (MoE 2013).

Collaboration is a core collectivist cultural value for Pāsifika people (MoE 2013). In this study, Pāsifika students valued group work because sharing ideas and strategies helped the students to progress and improve their own mathematics. Earlier research by Sharma et al. (2011) found that Pāsifika students recognised the benefit of collaborative mathematical learning both for building their own mathematical understanding and progressing their peers' mathematical understanding. Similarly, Anthony (2013) found that Pāsifika students valued a social arrangement in the classroom, which suited their collaborative ways of learning. This contrasts with research from Asia (e.g., Law et al. 2011; Zhang et al. 2016) which demonstrated an absence of collaborative mathematics values from East Asian students, and also research from Sweden (Seah and Peng 2012) that highlighted Swedish middle school students valuing independent working due to their perception that listening to their peers' conflicting strategies was confusing rather than helpful.

Effort/Practice

The mathematics educational value of effort/practice was explored through the statement: "*To be good at maths I need to practice with lots of questions*". This statement was ranked in the first three values of 24% ($n = 31/131$) of the students.

The key reason that students ($n = 19/31$) provided for the choice of this value indicated they viewed effort and practice as a way of facilitating their progress and achievement. Several students used the phrase "practice makes perfect" in relation to their mathematical learning. Specifically, a student stated: *If you practice you will get better and better.* Other students ($n = 6/31$) viewed this value as important as they saw effort and practice as a means of developing their conceptual understanding and clarity: *So you can understand the problems.* A small group of students ($n = 4$) linked the value of effort/practice to their future goals and success for education and employment. For example, one student stated: *So I don't have to struggle when it comes to a test or when I am in college*, another student explained: *So you can learn and you can get a better job.*

These responses indicated that students' perceived success and understanding in mathematics was achieved through hard work and practice. The results of the current study with Pāsifika students are broadly consistent with other values research from Europe and Asia (Lee and Seah 2015; Lim 2015; Österling and Andersson 2013) where students were reported to value effort and practice for their mathematics learning. This finding has important implications in challenging ongoing deficit perceptions and low expectations that many New Zealand teachers have towards their Pāsifika students (Rubie-Davies 2009, 2016). For example, Turner et al. (2015)

revealed New Zealand teacher expectations were highest for Asian and Pākehā (European) mathematics students, and lowest for Pāsifika and Māori, with one teacher in their study expressing that Pāsifika students were less likely to achieve in mathematics because they "are very lazy and they do not spend enough time studying and learning" (p. 62).

Family/Familial Support

The statement "*I cannot be good at maths without the support, love or guidance of my whānau/family*" was used as a value indicator for the mathematics educational value of family/familial support. Twenty-four percent of students ($n = 31/131$) ranked this statement in their first three values.

Most commonly, students ($n = 17/31$) described how their family (including extended family such as grandparents, or aunties and uncles) assisted with homework or taught them new skills or strategies. For example, a student stated: *If you have homework and you don't know how to do the strategy, then you can just ask your family*. The other key theme from the students ($n = 10/31$) for the reason of the importance of this value was the role of family in providing encouragement and support: *Every time I fail in maths, my Mum always encourages me to carry on and try my best.*

For these students, it was important that their family was actively involved in their mathematics learning. This finding is aligned to previous general education studies (e.g., Hannant 2013; J. Hunter et al. 2016) which found that values concerning family are central to the identity of Pāsifika learners with Pāsifika students frequently citing family as a major driver of their motivation and achievement. In the current study, the Pāsifika students valued their family as important for their mathematics learning because the family provided support with homework and learning along with encouragement. There has been limited mathematics education research investigating Pāsifika students' valuing of family support. The current finding is important to counteract the inaccurate stereotypes held by many New Zealand teachers relating to Pāsifika parents, that is that Pāsifika parents are not interested or involved in their children's schooling (Nakhid 2003), or that parents do not have the mathematical knowledge or skills to help with their children's homework (Nicholas and Fletcher 2015; Turner et al. 2015).

Conclusion and Implications

This study aimed to determine what Pāsifika students valued as most important for their mathematics learning. We discovered that Pāsifika students' valued mathematics which was useful and/or practical for the present or future, and valued effort and practice as important for success in mathematics. The commonality of these two values across international research studies (e.g., Barkatas and Seah 2015; Lee and Seah 2015; Lim 2015; Österling and Andersson 2013) suggests these mathematics educational values are influenced by common societal and educational values.

Interestingly, while identifying some values common with international studies, the findings from the current study also identify values specific to the local community, that is Pāsifika students in New Zealand schools. These were the values of peer collaboration-group work and family/familial support. This finding aligns with earlier research (e.g., Averill 2012; J. Hunter et al. 2016; R. Hunter and Anthony 2011) demonstrating a relationship between students' culture and their values within the mathematics classroom. Importantly, this finding highlights the need for educators to investigate the mathematics educational values of students from minority cultural groups, in order to both acknowledge and build on students' cultural backgrounds in the mathematics classroom.

Acknowledging students' values has important implications for culturally responsive, equitable and effective mathematics teaching, thus, it is important that teachers recognise what is being valued in their classrooms. For example, in the current study, the valuing of opportunities for group work and collaboration indicates pedagogical approaches that educators could adopt to align with their students' mathematics educational values. Furthermore, the valuing of effort and practice by Pasifika students offers a direct challenge to teachers' deficit and stereotypical views often reported in previous studies (e.g., Rubie-Davies 2009, 2016). This highlights the need for teachers, especially teachers of minority students, to consult with and determine what their students' value prior to making assumptions. As Seah (2016) contends by recognising the cultural uniqueness of what the students' value for their mathematics learning, teachers can customise instruction/activities to align with student values.

In the New Zealand context, it appears that there has been limited research exploring students' self-reported mathematics values. The findings from the current study may provide a useful starting point for developing an evidence base in New Zealand related to Pāsifika students as well as contributing to the international mathematics values literature. As Seah (2016) writes "how do we go about facilitating students' appropriate valuing such that it helps them to study mathematics more effectively? The first step may be to have a good idea of what is currently being valued by students" (p. 4). By recognising what is valued (or not valued) in the mathematics classroom, teachers can develop classroom culture or pedagogy which aligns with the students' values or work with students to examine values which may contradict the classroom norms and pedagogy. In alignment with previous studies (e.g., J. Hunter et al. 2016; Seah et al. 2016), we argue that when values are acknowledged in the mathematics classroom, relationships are strengthened, students' cultural identities are affirmed, students become more engaged, and ultimately, mathematics learning is enhanced.

References

Anthony, G. (2013). Student perceptions of the 'good' teacher and 'good' learner in New Zealand classrooms. In G. Anthony, B. Kaur, M. Ohtani, & D. Clarke (Eds.), *Student voice in mathematics classrooms around the world* (pp. 209–225). Rotterdam, Netherlands: Sense Publishers.

Attard, C. (2011). "My favourite subject is maths. For some reason no-one really agrees with me": Student perspectives of mathematics teaching and learning in the upper primary classroom. *Mathematics Education Research Journal, 23*(3), 363–377.

Averill, R. (2012). Caring teaching practices in multiethnic mathematics classrooms: Attending to health and well-being. *Mathematics Education Research Journal, 24*(2), 105–128.

Barkatsas, T., & Seah, W. (2015). Learners' preferred mathematical task types: The values perspective. In A. Bishop, H. Tan, & T. Barkatsas (Eds.), *Diversity in mathematics education* (pp. 63–79). Cham: Springer.

Bishop, A. (1996). How should mathematics teaching in modern societies relate to cultural values—some preliminary questions. In *Paper Presented at the Meeting of the Seventh Southeast Asian Conference on Mathematics Education, Hanoi, Vietnam.*

Caygill, R., Hanlar, V., & Singh, S. (2016a). *TIMSS 2015: New Zealand Year 5 Maths results.* Wellington: Comparative Education Research Unit, Ministry of Education.

Caygill, R., Hanlar, V., & Singh, S. (2016b). *TIMSS 2015: New Zealand Year 9 Maths results.* Wellington: Comparative Education Research Unit, Ministry of Education.

Civil, M., & Hunter, R. (2015). Participation of non-dominant students in argumentation in the mathematics classroom. *Intercultural Journal, 26*(4), 296–312.

Clarkson, P., Bishop, A. J., FitzSimons, G., & Seah, W. T. (2000). Challenges and constraints in researching values. In M. Goos, R. Brown, & K. Makar (Eds.), *Proceedings of the 31st annual conference of the Mathematics Education Research Group of Australasia* (pp. 188–195). Brisbane: MERGA.

Coxon, E., Anae, M., Mara, D., Wendt-Samu, T., & Finau, C. (2002). *Literature review on Pacific education issues*. Wellington: Ministry of Education.

Education Assessment Research Unit and New Zealand Council for Educational Research. (2015). *National monitoring study of student achievement: Mathematics and statistics 2013*. Wellington: Ministry of Education.

Gay, G. (2010). *Culturally responsive teaching: Theory, research, and practice* (2nd ed.). New York: Teachers College Press.

Gravemeijer, K., Stephan, M., Julie, C., Lai, F. L., & Ohtani, M. (2017). What mathematics education may prepare students for the society of the future? *International Journal of Science and Mathematics Education, 15,* 105–123.

Grootenboer, P., & Marshman, M. (2015). *Mathematics, affect and learning: Middle school students' beliefs and attitudes about mathematics education.* London: Springer.

Halstead, J. M. (1996). Values and values education in schools. In J. Halstead & M. Taylor (Eds.), *Values in education and education in values* (pp. 3–14). London: The Falmer Press.

Hannant, B. (2013). *What works: Academically successful Pāsifika males identify factors contributing to their educational outcomes (Unpublished Master's thesis).* Auckland: Massey Univeristy.

Hāwera, N., Taylor, M., Young-Loveridge, J., & Sharma, S. (2007). "Who helps me learn mathematics, and how?" Màori children's perspectives. In D. Holton (Ed.), *Findings from the New Zealand Numeracy Development Projects 2006* (pp. 54–66). Wellington: Ministry of Education.

Hill, J. (2017). *What do culturally diverse middle school students' value for their mathematics learning (Unpublished Master's thesis).* Auckland: Massey Univeristy.

Hunter, J., Hunter, R., Bills, T., Cheung, I., Hannant, B., Kritesh, K., et al. (2016). Developing equity for Pāsifika learners within a New Zealand context: Attending to culture and values. *New Zealand Journal of Educational Studies, 51*(2), 197–209.

Hunter, R., & Anthony, G. (2011). Forging mathematical relationships in inquiry-based classrooms with Pāsifika students. *Journal of Urban Mathematics Education, 4*(1), 98–119.

Hunter, R., Hunter, J. & Bills, T. (in press). Enacting culturally responsive or socially-response-able mathematics education. In C. Nicol, S. Dawson, J. Archibald & F. Glanfield (Eds.). *Living culturally responsive mathematics curriculum and pedagogy: Making a difference with/in indigenous communities*. Sense.

Kazemi, E., Franke, M., & Lampert, M. (2009). Developing pedagogies in teacher education to support novice teachers' ability to enact ambitious instruction. In. R. Hunter, B. Bicknell & T. Burgess (Eds.), *Proceedings of the 32nd Annual Conference of the Mathematics Education Research Group of Australalasia* (pp. 12–30). Palmerston North: MERGA.

Ladson-Billings, G. (1994). *The dreamkeepers*. San Francisco: Jossey-Bass Publishing.

Law, H. Y., Wong, N. Y., & Lee, N. Y. L. (2011). The Third Wave studies of values in effective mathematics education: Developing students' mathematical autonomy in classroom learning. *The Mathematics Educator, 13*(1), 72–86.

Lee, H. F., & Seah, W. T. (2015). *"Math is not for us, not an indigenous thing, you know": Empowering Taiwanese indigenous learners of mathematics through the values approach.* In *Paper Presented at the Eight International Mathematics Education and Society Conference, Portland, OR.*

Lim, C. S. (2015). *Riding the third Wave: Negotiating teacher and Students' value preferences relating to effective mathematics lessons.* In *Paper presented at the Selected Regular Lectures from the 12th International Congress on Mathematical Education, Seoul, Korea.*

Ministry of Business Innovation & Employment. (2017). *Economic development: Digital economy*. Wellington: Ministry of Business Innovation & Employment.

Ministry of Education. (2006a). *Education counts: School rolls*. Wellington: Ministry of Education.

Ministry of Education. (2006b). *Pāsifika Education Plan 2006–2010*. Wellington: Ministry of Education.

Ministry of Education. (2009). *Pāsifika Education Plan 2009–2012*. Wellington: Ministry of Education.

Ministry of Education. (2013). *Pāsifika Education Plan 2013–2017*. Wellington: Ministry of Education.

Nakhid, C. (2003). "Intercultural" perceptions, academic achievement, and the identifying process of Pacific Islands students in New Zealand schools. *Journal of Negro Education, 72*(3), 297–317.

Nicholas, K., & Fletcher, J. (2015). What supports 11-to 13-year-old Pāsifika students in mathematics learning in New Zealand classrooms? *Education, 45*(1), 1–15.

Österling, L., & Andersson, A. (2013). Measuring immeasurable values. In A. Lindmeier & A. Heinze (Eds.), *Proceedings of the meeting of the 37th Conference of the International Group for Psychology of Mathematics Education* (pp. 17–24). Kiel: PME.

Rubie-Davies, C. (2009). Teacher expectations and labeling. In L. Saha & A. Dworkin (Eds.), *International handbook of research on teachers and teaching* (pp. 695–707). Boston: Springer.

Rubie-Davies, C. (2016). High and low expectation teachers: The importance of the teacher factor. In S. Trusz & P. Babel (Eds.), *Interpersonal and intrapersonal expectancies* (pp. 145–157). Abington, Oxon: Routledge.

Seah, W. T. (2016). Values in the mathematics classroom: Supporting cognitive and affective pedagogical ideas. *Pedagogical Research, 1*(2), 45–63.

Seah, W. T., & Andersson, A. (2015). Valuing diversity in mathematics pedagogy through the volitional nature and alignment of values. In A. Bishop (Ed.), *Diversity in mathematics education* (pp. 167–183). Cham: Springer.

Seah, W. T., Andersson, A., Bishop, A., & Clarkson, P. (2016). What would the mathematics curriculum look like if values were the focus? *For the Learning of Mathematics, 36*(1), 14–20.

Seah, W. T., & Peng, A. (2012). What students outside Asia value in effective mathematics lessons: A scoping study. *ZDM: The International Journal on Mathematics Education, 44*(1), 71–82.

Seah, W. T., & Wong, N. Y. (2012). What students value in effective mathematics learning: A 'Third Wave Project' research study. *ZDM: The International Journal on Mathematics Education, 44*(1), 33–43.

Sharma, S., Young-Loveridge, J., Taylor, M., & Häwera, N. (2011). The views of Pāsifika students in New Zealand about communicating mathematically. *Asia Pacific Journal of Education, 31*(4), 503–519.

Turner, H., Rubie-Davies, C. M., & Webber, M. (2015). Teacher expectations, ethnicity and the achievement gap. *New Zealand Journal of Educational Studies, 50*(1), 55–69.

Young-Loveridge, J., Taylor, M., Sharma, S., & Hāwera, N. (2006). Students' perspectives on the nature of mathematics. In P. Grootenboer, R. Zevenbergen & M. Chinnappan (Eds.), *Proceedings of the 29th Annual Conference of the Mathematics Education Research Group of Australasia* (pp. 583–590). Adelaide: MERGA.

Zhang, Q., Barkatsas, T., Law, H. Y., Leu, Y. C., Seah, W. T., & Wong, N. Y. (2016). What primary students in the Chinese mainland, Hong Kong and Taiwan value in mathematics learning: A comparative analysis. *International Journal of Science and Mathematics Education, 14*(5), 907–924.

3

Why Mathematics is Valuable for Turkish, Turkish Immigrant and German Students? A Cross-Cultural Study

Yüksel Dede

Abstract This study investigated students' views regarding the value of mathematics. It reports on a smaller part of findings from a wider comparative study concerning student values of mathematics and mathematics education that belonged to Turkish students in Turkey, Turkish immigrant students in Germany, and German students in Germany. Students were in Grade 9 (14–15 years old) and the data was gathered through semi-structured interviews, and analyzed using the constant comparative method. The results revealed four major value categories for Turkish (practice, relevance, rationalism, and fun values) and German (relevance, fun, rationalism, and consolidating values) students while there were three major value categories for the immigrant students (relevance, rationalism, and communication).

Keywords Cross-cultural study · Turkish students · German students · Turkish immigrant students · Grade 9 students · Values · Mathematical values

Introduction

The process of values education occurs within a complex structure of the human interactions -such as learning, personal development, socializing and cognition- through the agencies of traditional customs, norms, and language. In other words, these interactions occur in the context of culture. Value systems are therefore an integral part of any cultural context (Thomas 2000). Although there is no consensus about the definition of the concept 'culture', often people have a general understanding of what culture is and what it requires. In this regard, culture consists of values, beliefs, and concepts that are shared within a society (Venaik and Brewer 2008). In this context, we examine and compare mathematics values of students living in different cultures (Turkish students in Turkey, Turkish immigrant students in Germany, and German students in Germany).

Y. Dede (✉)
Gazi Education Faculty, Department of Mathematics and Science Education, Gazi University, Ankara, Turkey
e-mail: ydede@gazi.edu.tr

Theoretical Background

This section discuss the relationships between mathematics and values, as well as learning about values through comparative research.

Values and Mathematics

Nowadays, although values are important to researchers and educators, the concept of values is elusive and broad, therefore the discussion of values can be found in most disciplines. The definitions of values can be extended from personal to collective levels and to many forms of knowledge (Lee and Manzon 2014). Indeed, the word "value" is used in different contexts for different meanings (see Seah and Bishop 2000). However, in general, values are general guides for the behavior emerging from one's experiences and relations (Raths et al. 1987). From this point of view, values play a role in one's choices, decisions, and behaviors consciously or unconsciously (FitzSimons et al. 2001). Seah (2003) also saw value as "an individual's internalization, 'cognitisation' and decontextualization of affective constructs (such as attitudes and beliefs) in his/her socio-cultural context" (p. 2). As Ernest (2009) stated,

> Mathematics is viewed as value-neutral, concerned only with structures, processes and the relationships of ideal objects, which can be described in purely logical language. ... In contrast, the (fallibilist) view of the new philosophy of mathematics is that the cultural values, preferences, and interests of the social groups involved in the formation, elaboration, and validation of mathematical knowledge cannot be so easily factored out and discounted (p. 57).

These two different points of view described by Ernest, related to mathematical philosophy, have quite different effects on how classroom practices are understood (see Ernest 1991). In this chapter we take the second perspective.

The USA National Council of Teachers of Mathematics [NCTM] (2000) regards mathematics as a part of the cultural heritage and describes it as one of the most important cultural and intellectual accomplishments of the human brain. Prediger (2001) characterizes mathematics as a "cultural phenomenon" (p. 23). Bishop (2001) also declared the importance of values (in mathematics education) as follows:

> Values exist on all levels of human relationships. On the individual level, learners have their own preferences and abilities that predispose them to value certain activities more than others. In the classroom, values are inherent in the negotiation of meanings between teacher and students and among the students themselves. At the institutional level, we enter the political world. Here, members of organizations engage in debates about both deep and superficial issues, including priorities in determining local curricula, schedules, teaching approaches, and so on. The larger political scene is at the societal level, where powerful institutions determine national and state priorities for mathematics curricula, teacher-preparation requirements, and other issues. Finally, at the cultural level, the very sources of knowledge, beliefs, and language influence our values in mathematics education (p. 347).

Values are part of educational processes and one important aspect to the conative environment of mathematics teaching. Not only mathematical knowledge but also mathematical values are consciously and at times unconsciously learnt by students. So, it is important for teachers and their students to be aware of the values they hold and to develop an awareness of values and value preferences toward teaching and learning respectively (Chin 2006). Students' accepted values play important roles (positive or negative) in their adult lives and professional workplaces (see Bishop et al. 2001; Dawkins and Weber 2016; Rhodes and Roux 2004).

Learning About Values Through Comparative Studies

In this chapter we take the position that culture is a powerful determiner of mathematical values. We also acknowledge that different cultures possess different values (Seah 2003) and investigating different cultures might help us to understand the nature and diversity of our own value systems. In this manner, too, school education in one country can be better understood in comparison to education in other countries. Moreover, international comparative studies can not only provide data on diagnosing and making decisions about students' learning, they are also able to shed light on issues relating to education in general and learning and teaching mathematics in particular (Cai 2006). Crossley and Watson (2003) also discussed some benefits of conducting comparative studies.

In this sense, this study makes a contribution towards what we can learn from comparative studies. In particular, this chapter documents a small part of a larger comparative study of Turkish students, Turkish immigrant students in Germany, and German students regarding their values towards mathematics and mathematics education. They were asked why mathematics is valuable, and the underlying mathematical values were explored.

Turkey and Germany had been selected to be compared against each other in this study, because they are two nations with huge cultural differences between them. The Federal Republic of Germany is an example of Western, liberal culture and has a multicultural society. Turkey is often seen as a bridge between Western and Eastern cultures, and although it has taken a series of steps towards Westernization, Turkey is still quite different from Germany with regards to culture, language and religion in particular.

Therefore, this study has the potential to provide an explanation to better understand how the different groups of teenagers' values regarding mathematics are similar or different. Also, the results might constitute a rich resource of ideas for future studies in educational development. It is also suggested that cultural differences—with the different underlying values—may influence how the same mathematical content might be taught through different approaches and different assessment emphases (Seah 2003).

In addition, Turkey and Germany are two different countries in terms of educational systems and mathematics education. The Turkish Ministry of National Education [MEB] is responsible for compulsory education in Turkey. Compulsory education in Turkey is free and it was first extended from 5 to 8 years in 1997 and later extended to 12 years in 2012, being implemented in the 2012–2013 academic year. The first four years of compulsory education are called elementary school, the second four years are called middle school, and the last four years are called high school. In other words, 7–10 year-old students generally attend elementary school, 11–14 year-old students generally attend middle school, and 15–18 year-old students generally attend high school. The Turkish education system is focused on high stake examinations with multiple-choice tests that are taken at the end of middle and high school. This situation causes a lot of pressure on the students. The results of large-scale national assessments (e.g. The University Entrance Exams), and international comparative studies (e.g. The Program for International Student Assessment [PISA] and The Trends in International Mathematics and Science Study [TIMSS]) report that Turkey achieved under-average mathematics scores (see MEB 2013, 2016). In order to improve performance, mathematics curricula reforms have been revised in Turkey. The curricula for both primary and secondary schools were first updated in 2005 and then revised again in 2013 and 2018. The purpose of the revisions was to change the mathematics curricula so that it would focus on learner-centered teaching and multidisciplinary approaches.

On the other hand, responsibility for organizing the education system is shared among 16 Länder (states) and the federal government in Germany. Unless Grundgesetz awards legislative powers to the Federation, the Länder have the right to legislate. Länder have their own education ministries and are responsible for schools, higher education, adult education and continuing education. Co-ordination between them is ensured by several bodies. The Standing Conference of the Ministers of Education and Cultural Affairs of the Länder in the Federal Republic of Germany (Kultusministerkonferenz [KMK]) co-ordinates education policies and makes recommendations for further developments in the area of primary and secondary education, higher education, research and cultural policy. Education is compulsory from ages 6 to 18 in Germany. System-level policies such as tracking, grade repetition, and academic selection can still hinder equity, especially for students with an immigrant background. PISA 2012 results indicated that these students' mathematics scores were 25 points lower than that of native students. Tracking begins at an early age in most Länder, and some Länder have strategies to limit its potentially negative effects on equity. In Hessen, for example, students can choose between 4- and 6-year primary schools, and in Berlin and Brandenburg, all primary schools are comprehensive until grade 6 (age 12) (OECD 2014). Grades 5 and 6 constitute a phase of particular promotion, supervision and orientation with regards to the pupil's future educational path and its particular direction. The general education qualifications that may be obtained after grades 9 and 10 carry particular designations in some Länder. Admission to the Gymnasiale Oberstufe requires a formal entrance qualification which can be obtained after grade 9 or 10. Since 2012, in the majority of Länder the Allgemeine Hochschulreife (Abitur-high school diploma) can be obtained after the successful completion of 12 consecutive school years (eight years at the Gymnasium) (KMK 2017). Although German students' mathematics scores in PISA 2000 were considered to be poor in comparison to some Asian and European countries (Misek 2007; Schumann 2000), German students achieved above-average mathematics scores in PISA 2012. That also revealed a significant improvement in their performance since 2000 (OECD 2014). The results of PISA 2015 showed that German students' average score for mathematics decreased compared to the results of PISA 2012 (OECD 2016).

Given the different trends in the two countries' education systems and achievements in school mathematics in international comparative exercises, it is thought that an understanding of mathematical values of students from these two countries may contribute to relevant literature. For example, with the changes in performance, have there been corresponding changes in what the students value in their mathematics learning?

Thirdly, it is expected that the comparisons made would provide a significant contribution to the literature and discussion concerning which values may be learnt by immigrant students, in this case Turkish students in Germany. The majority of students in Germany mostly come from a multicultural background, whereas Turkish students usually come from a homogeneous background.

Although there are some studies in German schools that examine the skills of German/Turkish bilingual students regarding their language use in doing mathematics (Schüler-Meyer et al. 2017), not many specifically investigate these students' values towards mathematics and mathematics education.

As such, the research questions posed in this study were,

(1) Is mathematics valuable to Turkish students, Turkish immigrant students in Germany, and German students?
(2) If so, how are the values associated with such positions different and similar for Turkish students, Turkish immigrant students in Germany, and German students?

Methodology

Research Design

This data in this study was obtained from the 'Students Values in Mathematics Teaching in Germany and Turkey' [SVMGT] project, which took place over a one-year period in 2015/2016. The objectives of the SVMGT project were to document values of Turkish students, Turkish immigrant students living in Germany, and German students regarding mathematics and mathematics learning and to explore the cultural, social and connected nature of these students' values. The SVMGT project adopted a sequential mixed method research design in which quantitative and qualitative research methods are used together (see Creswell and Clark 2007). In this project, the first phase of this method was the quantitative data collection (with the translation and adaption of the international survey questionnaire within The Third Wave Project, "What I Find Important" (WIFI) project) and analysis. The results of this phase will be reported as a separate study. In the second phase, based on the findings of the quantitative data, we gathered and analyzed qualitative data by means of semi-structured interviews. The aim has been to understand and compare the students' values towards mathematics and mathematics education.

Participants

Participants were chosen using the maximum variation sampling method. The participants of the study were 11 Turkish immigrant students living in Germany, 14 German and 10 Turkish students, all from Grade 9. All German students and Turkish immigrant students in Germany attended secondary schools (4 Gymnasiums and 7 high schools with average achieving students) in a province in northern Germany. Turkish students, on the other hand, attended secondary schools (4 Anatolian high schools with average and higher achieving students) in a province in the Central Anatolia Region.

Semi-structured Interviews

The interviews were carried out in a comfortable and an appropriate location by the researcher. The interviews were audio taped after obtaining the permission of each interviewee. Each interviewee was given an ID code (T1, … for Turkish students, TG1, … for Turkish immigrant students living in Germany, and G1, … for German students). Each interview lasted about 10–25 min. The interviews were conducted in students' native languages except the Turkish immigrant students- they were interviewed in either Turkish or German according to their preferences. Hence language issues for students were minimized. The interviews with German students were translated to Turkish by the researcher. The same interviews were also independently translated into Turkish by two college students who were able to speak Turkish and German at an advanced level and were enrolled in educational sciences in a German university. After all translations were completed, the translated documents were compared with regards to differences and similarities in order to enable utmost agreement among translations.

Data Analysis

We analyzed the data collected from the semi-structured interviews by using the constant comparative method (Strauss and Corbin 1998). The analysis of the data collected in the study was continued until the saturation was reached (Arber 1993). It was assumed that students might not be able to relate to values directly, so the questions in the interviews were about different learning activities that would be regarded as value indicators. This enabled the researcher to reflect on the problem of marking a difference between a value and a value indicator (Andersson and Österling 2013). For example, the learning activity "connecting mathematics to real life" in the semi-structured interviews was categorized as an indicator of the value of relevance. In this regard, a German student (G10) (Grade 9, 14 age, and mathematics score: 1–2 which represent the high levels in Germany) responded to the question from the interviewer:

> I:Is mathematics valuable for you?
>
> G10:Yes, it is so valuable
>
> I:Why?
>
> G10:I find it fun. It's easy for me since primary school. For this reason, I have no a problem in mathematics ... (pause). It is absolutely important. I will need math in my future career. I want to study engineering and so I will have to solve complex problems. I can solve them with mathematics. I care so much now that it's important for my job ... (pause). In that sense, it's the only reason. And I find math easy. That's why I love mathematics.

For G10, the statements of "I find it fun" and "That's why I love mathematics" both correspond to the value of *fun*. On the other hand, the statement "I will need math in my future career...I want to study engineering..." corresponds to the value of relevance. Transcripts of all the interviews were analyzed using the same coding process.

Trustworthiness

The categories emerged in this study were compared with Lim and Ernest's (1997) category of values taught in mathematics lessons, Bishop's (1988) category of mathematical values, and Hofstede's (2009) category of cultural values so that "theoretical triangulation" (Cohen et al. 2007, p. 142) was performed on the categories. In order to categorize the data gathered from semi-structured interviews and to identify common expressions, interview transcripts were read several times. Students' expressions were transcribed without any changes and these verbatim transcripts were submitted to the approval of the students, which provided "member check" (Creswell 1998) for the reliability of the interview data. "Peer review" was also applied for the reliability of the research data. According to Lincoln and Guba (1985), peer review is an external control mechanism for the research reliability. Thus, major and sub-categories created by the researcher were sent to two separate researchers—one of them had a Ph.D. degree in mathematics education and the other had a Ph.D. degree in science education. According to the expert opinions, sub-categories were revised.

Results and Discussion

The results revealed that there were 9 different value indicators (preparing for examinations tests, mathematics in daily life, relationships to other subjects in school, future career, and understand real-world) and 4 corresponding values (practice, relevance, rationalism, and fun) for the Turkish students, 7 different value indicators (mathematics in daily life, relationships between mathematics concepts, future career, calculation, reasoning, universal language) and 3 corresponding values (relevance, rationalism, and communication) for the Turkish immigrant students, and 14 different value indicators (mathematics in daily life, applicability, relationships between mathematics concepts, relationships to other subjects in school, understand real-world, future career, game, structural, reasoning, systematic, precise, calculation, visualization, and concretization) and 4 corresponding values (relevance, fun, rationalism, consolidating) for the German students. Out of these values, the value of *rationalism* pertains to the mathematical values of Bishop (1988) or epistemological values of Lim and Ernest (1997), whereas the values of relevance, fun, practice, consolidating, and communication pertain to mathematics educational values of Bishop (1988) and to social and cultural values of Lim and Ernest (1997) to some extent. The value indicators and their corresponding values were also categorized into two types of thinking: isolated thinking and connected thinking. Isolated thinking reflects that mathematics is seen as a set of isolated concepts and procedures whereas connected thinking reflects that connected values such as connections among mathematical ideas and ideas from other disciplines usefulness, process, communication, and creativity (see Dede 2012; Ernest 2004). Description of value indicators, their corresponding values, and the types of thinking are summarized in Table.

As can be seen from Table again, 14 value indicators (five of them are related to isolated thinking and nine to connected thinking) for German students, 9 value indicators (three of them are related to isolated thinking and six to connected thinking) for Turkish students, and 7 value indicators (two of them are related to isolated thinking and five to connected thinking) for the immigrant students emerged from the interviews. These findings indicated that German students identified a wider range of values than the Turkish and the immigrant students for both of the value indicators. And it also reflected that all groups of students put more emphasis on the connected values. Moreover, the findings indicated that there were similarities and differences (discussed below) among Turkish, Turkish immigrant, and German students' value indicators and their corresponding values regarding to why mathematics is valuable.

Similarities

As can be seen from Table, four value indicators (mathematics in daily life, relationships to other subjects in school, future career and reasoning) are common among three groups of students. Common values across the three groups of students

Table: Comparison of the students' value indicators, corresponding values and types of thinking

Value indicator (corresponding value)*			
Type of thinking	Turkish	Turkish immigrant students in Germany	German
Connected thinking	1. Preparing for examinations (high stake tests), tests (Practice) 2. Mathematics in daily life (Relevance) 3. Relationships to other subjects in school (Relevance) 4. Future career (Relevance) 5. Understand real-world (exp. golden ratio) (Relevance) 6. Puzzle, game (Fun)	1. Mathematics in daily life (Relevance) 2. Relationships between mathematics concepts (Relevance) 3. Relationships to other subjects in school (Relevance) 4. Future career (Relevance) 5. Universal language (Communication)	1. Mathematics in daily life (Relevance) 2. Applicability (Relevance) 3. Relationships between mathematics concepts (Relevance) 4. Relationships to other subjects in school (Relevance) 5. Understand real-world (Relevance) 6. Future career (Relevance) 7. Game, puzzle (Fun) 8. Visualization (Consolidating) 9. Concretization (Consolidating)
Isolated thinking	1. Reasoning (Rationalism) 2. Precise (Rationalism) 3. Clear (Rationalism)	1. Calculation (Rationalism) 2. Reasoning (Rationalism)	1. Structural (Rationalism) 2. Reasoning (Rationalism) 3. Systematic (Rationalism) 4. Precise (Rationalism) 5. Calculation (Rationalism)

*The corresponding value is given in parentheses

are rationalism and relevance. The value indicators corresponding to these two values indicate that German students offer a larger variety of value indicators than the other groups for both rationality and relevance values. On the other hand, fun was a common value for only the German and Turkish students. As mentioned above, these results indicated that all three groups of students put emphasis on the values of relevance and rationalism. Similar results for the student valuing of relevance had been reported for the three Chinese regions of Chinese Mainland, Hong Kong and Taiwan (Zhang et al. 2016). Equally significantly, both these values (i.e. relevance and rationalism) were also found to be embraced by both Turkish and German teachers (Dede 2012). Indeed, Australian primary school teachers were also observed to subscribe to similar values (Bishop et al. 2001). This might suggest that mathematics teachers in different cultures, like the three groups of students in the present study, hold common approaches with regard to the values related to the scientific discipline of mathematics (e.g. rationalism) (Atweh and Seah 2008). Also, these results are in line with the Platonist view of mathematics, which is not surprising for the German students. Kaiser and Vollstedt (2007) suggest that "the Gymnasium shows a strong dominance of theoretical subject-related reflections." (p. 346). The values of *relevance* and *fun* are also generally consistent with the objectives and expectations of the primary and secondary I and secondary II level mathematics curricula (see Rahmenplan Grundschule Mathematik [RGM], Rahmenlehrplan für die Sekundarstufe-I [RSS-I], Rahmenlehrplan für die Gymnasiale Oberstufe [RGO]). At the same time, this result is in line with the fallibilist view of mathematics. For example, RSS-I in Germany (Berlin) consists of the following principles (RSS 2006):

> Mathematics is a science which can be applied to several areas. It allows capturing and also solving the math structures and problems from both science and technical and real life … Mathematics, in this way, develops methods and examines objects and opinions. Mathematics encourages improving humanistic thinking, creativity and problem solving skills in science and real life (p. 9).

On the other hand, the findings about *practice, relevance* and *rationalism* values related to the Turkish sample are generally consistent with the objectives of the Turkish curriculum (MEB 2005):

> Learning and teaching the system of mathematical thinking; relating basic mathematical skills (e.g. problem solving, reasoning, connections, generalization, and affective and psycho-motor skills development) and abilities based on these skills to real life problems; improving their mathematics skills and abilities while preparing youth for real life through math studies;…understanding some of the elements on which math is based; assessing our place in earth, culture and society; teaching the importance of math in the artistic dimension; teaching that math is systematic knowledge and a computer language; … (pp. 4–5).

Differences

As can be seen from Table, consolidating appeared to be valued by the German students only, whereas practice was a value that was embraced by only the Turkish students. Similarly, communication was a value for only the immigrant students. And while the immigrant students valued *rationalism* just like their German peers and fellow Turkish peers in Turkish schools, their valuing was associated with only two value indicators, namely, calculation and reasoning. It is also interesting that the fun value was not found in the immigrant students' statements. The valuing of practice may reflect the Turkish education system's focus on high-stakes examinations as indicated in the literature section. With these exams, Turkish students' mathematical skills are assessed as well as their ability to use time in the most efficient manner possible. That means, if students want to succeed in these exams, they should solve a lot of mathematical questions and problems. Similarly, the result for communication value in the immigrant students may be related to several factors such as integration, language barriers and multilingualism. Besides that, communication value can be considered within the collectivism dimension of Hofstede's cultural values (2009) as well as societal values (see Dede 2013).

Moving On

This study has provided evidence that the values for the nature of mathematics (rationalism) and the way of teaching (relevance) are common for the students in two different cultures. The study also showed evidence that mathematics could be a tool for the immigrant students to communicate with the culture they live in. Also, the study has pointed out evidence that values across the three groups of students generally consistent with the objectives of their mathematics curriculum.

It has also demonstrated how comparative studies can help us to understand education systems, teachers' work, and students' learning in ways which are not evident or possible with studies drawing their data from one culture only. Indeed, the results of the current study have revealed interesting similarities and differences in terms of students' mathematical values in Germany and Turkey. These similarities and differences would not have been as visible in a study involving German students or Turkish students alone.

Questions regarding the causes and impacts of these similarities and differences may also be identified. For example, further research may be carried out to elaborate the reasons underlying the results for fun and communication values that came from immigrant

Turkish students as well as for consolidating value held by the German students and for practice value held by the Turkish students. Moreover, further research focusing on the values of communication and fun may be carried out based on the concepts of Bishop's proposed enculturation (1988) and acculturation (2002).

As mentioned earlier, the study is limited with the students' views to the question of "why mathematics is valuable?" in two different countries. For this sense, a further study could employ classroom observations and in-depth interviews with more questions with the students in both countries in order to explain how the causes and impacts of the similarities and differences came about. Moreover, further research could examine students' mathematical values in different cultures for different mathematics contexts (e.g. preparing for mathematics lessons, learning methods, and decision-making process etc.). By doing so, more information on students' mathematical values can be collected. Finally, due to the small sample size, it is difficult to generalize the findings of this present study to other settings. Further research could examine whether similar results can be obtained from a study with a larger sample.

Acknowledgements The study involving students from Turkey was supported, in part, by grants from the Scientific and Technological Research Council of Turkey (TUBITAK, Grant No: 1059B191401095). Any opinions expressed herein are those of the author and do not necessarily represent the views of TUBITAK. Special thanks to Prof. Dr. Uwe Gellert, Free University of Berlin, Germany.

References

Andersson, A., & Österling, L. (2013, July–August). Measuring immeasurable values. In A. M. Lindmeier & A. Heinze (Eds.), *Paper presented at the Proceedings of the 37th Conference of the PME* (Vol. 2, pp. 17–24). Kiel, Germany.

Arber, S. (1993). Designing samples. In N. Gilbert (Ed.), *Researching social life* (pp. 68–92). London: Sage.

Atweh, B., & Seah, W. T. (2008). Theorising values and their study in mathematics education. *Presented at the AARE Conference*. Fremantle, Australia. http://www.aare.edu.au/07pap/atw07578.pdf.

Bishop, A. J. (1988). *Mathematical enculturation: A cultural perspective on mathematics education*. Dordrecht, The Netherlands: Kluwer Academic Publishers.

Bishop, A. J. (2001). Educating student teachers about values in mathematics education. In F. L. Lin & T. Cooney (Eds.), *Making sense of mathematics teacher education* (pp. 233–246). Holland: Kluwer Academic Publishers.

Bishop, A. J. (2002). Mathematical acculturation, cultural conflicts, and transition. In G. de Abreu, A. J. Bishop, & N. C. Presmeg (Eds.), *Transitions between contexts of mathematical practices* (pp. 193–212). Dordrecht, Holland: Kluwer.

Bishop, A., Clarkson, P., FitzSimons, G., & Seah, W. T. (2001). Studying values in mathematics education: Aspects of the VAMP Project. In J. Novotna (Ed.), *European research in mathematics education II* (pp. 368–376). Prague: Charles University.

Cai, J. (2006). U.S. and Chinese teachers' cultural values of representations in mathematics education. In F. K. S. Leung, K. D. Graf, & F. Lopez-Real (Eds.), *Mathematics education in different cultural traditions: A comparative study of East Asian and the West* (pp. 465–482). New York, NY: Springer.

Chin, C. (2006). Conceptualising pedagogical values and identities in teacher development: A comparison of Taiwanese and Australian mathematics teachers. In F. K. S. Leung, K. D. Graf, & F. J. Lopez-Real (Eds.), *Mathematics education in different cultural traditions: A comparative study of East Asia and the West* (pp. 537–547). NY: Springer.

Cohen, L., Manion, L., & Morrison, K. (2007). *Research methods in education* (6th ed.). London: Taylor & Francis Group, Routledge Falmer.

Creswell, J. W. (1998). *Qualitative inquiry and research design: Choosing among five traditions*. Thousand Oaks, CA: Sage.

Creswell, J., & Clark, V. (2007). *Designing and conducting: Mixed methods research*. Thousand Oaks, CA: Sage.

Crossley, M., & Watson, K. (2003). *Comparative and international research in education: Globalisation, context and difference*. London: Routledge Falmer.

Dawkins, P., & Weber, K. (2016). Values and norms of proof for mathematicians and students. *Educational Studies in Mathematics, 95,* 123–142.

Dede, Y. (2012). Why is mathematics valuable? A comparison of Turkish and German mathematics teachers. *Mathematics Education Bulletin—BOLEMA, 26*(44), 1171–1206.

Dede, Y. (2013). Examining the underlying values of Turkish and German mathematics teachers' decision making processes in group Studies. *Educational Sciences: Theory & Practice, 13*(1), 670–706.

Ernest, P. (1991). *The philosophy of mathematics education*. London: Falmer Press.
Ernest, P. (2004). Images of mathematics, values and gender: A philosophical perspective. In B. Allen & S. Johnston-Wilder (Eds.), *Mathematics education: Exploring the culture of learning* (pp. 11–25). London: Routledge.
Ernest, P. (2009). New philosophy of mathematics: Implications for mathematics education. In B. Greer, S. Mukhopadhyay, A. B. Powell, & S. Nelson-Barber (Eds.), *Culturally responsive mathematics education* (pp. 43–64). New York, NY: Routledge.
FitzSimons, G., Seah, W. T., Bishop, A. J., & Clarkson, P. (2001). *Beyond numeracy: Values in the mathematics classroom*. 24th Annual MERGA Conference, Sydney, Australia.
Hofstede, G. H. (2009). *Geert Hofstede's cultural dimensions*. http://www.geert-hofstede.com/hofstede_germany.shtml.
Kaiser, G., & Vollstedt, M. (2007). Teachers' views on effective mathematics teaching: Commentaries from a European perspective. *ZDM Mathematics Education, 39*(4), 341–348.
Kultusministerkonferenz [KMK]. (2017). *Basic structure of the education system in the Federal Republic of Germany*. Berlin: German EURYDICE Unit of the Länder.
Lee, W. O., & Manzon, M. (2014). Comparing values. In M. Bray, B. Adamson, & M. Mason (Eds.), *Comparative education research: Approaches and methods* (pp. 259–283). Hong Kong: Springer & Comparative Education Research Centre, the University of Hong Kong.
Lim, L., & Ernest, P. (1997, March). Values in mathematics education: What is planned and what is espoused? *Proceedings of the Day Conference: In British Society for Research into Learning Mathematics*. University of Nottingham.
Lincoln, Y. S., & Guba, E. G. (1985). *Naturalistic inquiry*. Newburry Park, CA: Sage.
Misek, M. (2007). *Germany's educational tracking system and how it affects entrepreneurship*. University of Illinois at Urbana-Champaign. http://www-sre.wu-wien.ac.at/neurus/Misek07.pdf.
National Council of Teachers of Mathematics [NCTM]. (2000). *Principles and standards for school mathematics*. Reston, VA: Author.
Prediger, S. (2001). Mathematik als kulturelles Produkt menschlicher Denktätigkeit und ihr Bezugzum Individuum. In K. Lengnink, S. Prediger, F. Siebel (Eds.), Mathematik und Mensch (pp. 21–36). *Sichtweisen der AllgemeinenMathematik*, Darmstädter Schriften zur Allgemeinen Wissenschaft 2, Verlag Allgemeine Wissenschaft, Mühltal.
Rahmenlehrplan für die Gymnasiale Oberstufe [RGO]. (2006). *Wissenschaft und Technik*. Verlag: Berlin, Germany.
Rahmenlehrplan für die Sekundarstufe-I [RSS]. (2006). *Wissenschaft und Technik*. Verlag: Berlin, Germany.
Rahmenplan Grundschule Mathematik[RGM]. (2004). *Wissenschaft und Technik*. Verlag: Berlin.
Raths, L. E., Harmin, M., & Simon, S. B. (1987). Selections from values and teaching. In P. F. Carbone (Ed.), *Value theory and education* (pp. 98–214). Malabar: Krieger.
Rhodes, B., & Roux, C. (2004). Identifying values and beliefs in an outcomes-based curriculum. *South African Journal of Education, 24*(1), 25–30.
Schüler-Meyer, A., Prediger, S., Kuzu, T., Wessel, L., & Redder, A. (2017). Is formal language proficiency in the home language required to profit from a bilingual teaching intervention in mathematics? A mixed methods study on fostering multilingual students' conceptual understanding. *International Journal of Science and Mathematics Education*, 1–23.
Schumann, H. (2000). The German educational system and math curriculum. In *Hong Kong Mathematics Education Conference* (pp. 1–108). Chinese University of Hong Kong. www.ph-weingarten.de/homepage/lehrende/schumann/hongkong/german.pps.
Seah, W. T., & Bishop, A. J. (2000). *Values in mathematics textbooks: A view through two Australasian regions*. Paper Presented at the Annual Meeting of the American Educational Research Association, New Orleans, LA.
Seah, W. T. (2003). *The professional socialisation of teachers in transition: A values perspective*. Paper presented at the International Education Research Conference AARE-NZARE, Auckland, New Zealand.

Strauss, A., & Corbin, J. (1998). *Basics of qualitative research: Grounded theory procedures and techniques* (2nd ed.). Newbury Park, CA: Sage.

The Organization for Economic Cooperation and Development [OECD]. (2014). *Education policy outlook*. Germany: OECD Publishing.

The Organization for Economic Cooperation and Development [OECD]. (2016). *PISA 2015 assessment and analytical framework: Science, reading, mathematics and financial literacy*. Paris: PISA, OECD Publishing.

Thomas, E. (2000). Researching values in cross-cultural contexts. In R. Gardner, J. Cairns, & D. Lawton (Eds.), *Education for values: Morals, ethics and citizenship in contemporary teaching* (pp. 259–274). London: Kogan Page Limited.

Turkish Ministry of National Education [MEB]. (2005). *High school mathematics curriculum, grade 9-12*. Ankara: National Education Press.

Turkish Ministry of National Education [MEB]. (2013). *PISA 2012, National prereport*. Ankara: National Education Press.

Turkish Ministry of National Education [MEB]. (2016). *PISA 2016, National Report*. Ankara: National Education Press.

Venaik, S., & Brewer, P. A. (2008, June–July). Contradictions in national culture: Hofstedevs GLOBE. In J. Cantwell & T. Kiyak (Eds.), *50th annual meeting of the academy of international business (AIB)* (p. 274). Milan, Italy.

Zhang, Q., Barkatsas, T., Law, H. Y., Leu, Y.-C., & Seah, W. T. (2016). What primary students in the Chinese Mainland, Hong Kong and Taiwan value in mathematics learning: A comparative analysis. *International Journal of Science and Mathematics Education, 14*(5), 907–924.

4

Values in Mathematics Learning: Perspectives of Chinese Mainland Primary and Secondary Students

Qiaoping Zhang

Abstract Using the What I Find Important (in mathematics learning) (WIFI) framework (Seah et al. 2017), this study investigated the aspects of mathematics learning that were ascribed particular value by the Chinese Mainland primary and secondary students. Compared with the secondary students, the primary students tended to attribute greater value to ability, effort, diligence, use of formulas and memory. The secondary students were more likely to value knowledge and thinking as components of mathematics learning. Students in general valued a teacher-led yet student-centered learning approach. In addition, gender differences were observed. Specifically, the boys tended to attribute greater value to ability, rational understanding and creativity than girls, whereas the girls tended to value mathematical exploration more highly.

Keywords Values in mathematics learning · WIFI study · The Chinese Mainland students

Introduction

Undoubtedly, providing quality education and effective teaching for citizens is important for all countries in the world. Ways of ensuring effective mathematics teaching and learning have received attention from researchers for decades. Whether a teaching method or strategy is effective and successful depends on how the teacher uses it and the teacher's judgment. Such judgments actually reflect the teachers' mathematics education values. In school mathematics, values are reflected through the mathematics curriculum, pedagogical practices and individual views on the relevance and importance of mathematics, mathematical activities, and the study of mathematics. But values are not just matters of individual decisions. Societal influences also play a role and hence differences between countries can also be observed: practicing

---

Q. Zhang (✉)
The Education University of Hong Kong, Tai Po, Hong Kong SAR, China
e-mail: zqiaoping@eduhk.hk

and memorizing are valued in the Eastern education system in general whereas communication and critical thinking are emphasized in the Western system.

Values in mathematics education are the convictions internalized by an individual to be the most important and worthwhile components of the learning and teaching of mathematics (Bishop 1988; Seah and Andersson 2015). Many studies have been influenced by Bishop's (1988) three pairs of complementary values; namely *rationalism* and *objectism*, *control* and *progress*, and *mystery* and *openness*. Later, he argued that the two values in each pair might appear to be opposites of each other, but are actually complementary and that all are fostered through mathematics learning at school (Bishop 1999, p. 2). In fact, the kind of values in mathematics that are recognized by students, is not only influenced by the mathematical knowledge presented in textbooks, but are also influenced by their teachers' characteristics and teaching style (Opdenakker and van Damme 2006, p. 16). These values are related to mathematics educational values which are in turn related to the norms and practice of mathematics pedagogy and show students what is required to learn mathematics well (Atweh and Seah 2008; Seah et al. 2017).

Previous Research

Research on values in mathematics education is mainly divided into three fields: teachers' values in mathematics teaching (Dede 2015; Bishop et al. 2001), values in mathematics curriculum or textbooks (Dede 2006; Seah et al. 2016) and students' values (Seah and Wong 2012) with the majority of studies dealing with teachers' values. Little knowledge is known about values from the students' perspective. Given the deeply affective qualities of values, much more research in this area is required, particularly empirical studies focussed on what students themselves value in their learning experiences. Rather than thinking of mathematics teaching as just teaching mathematics to students, we should remember that we are also teaching students through mathematics. They are learning values related to the subject through how they are being taught. What students see as valuable in their mathematics learning experience is worthy to know. Thus, the purpose of this current study was to investigate what the Chinese Mainland students value in their mathematics learning.

In the Third Wave Project (Seah and Wong 2012), pairs of key values relating to mathematics education were identified including *ability* and *effort*, *wellbeing* and *hardship*, *process* and *product*, and *application* and *computation*. These pairs of mathematics educational values were often identified in East-Asian mathematics classroom. In Hong Kong, junior students considered enjoyment, order, achievement, student involvement, teacher-led monitoring, and teacher support to be vital to the effective learning of mathematics (Law et al. 2012). Malaysian primary students valued board work, exercise or practice, learning through mistakes, explanation and students' involvement as the five common elements of an effective mathematics lesson (Lim 2015). In Japan, fifth grade students tended to value process, effort, exploration, fact, openness and progress, whereas ninth grade students tended to value product, ability, exposition, idea, mystery, and control (Shinno et al. 2014) suggesting students' values can change over time. Zhang et al. (2016) reported on primary students' values in mathematics learning from the Chinese Mainland, Hong Kong and Taiwan. It was found that six value components formed the value structure. They were achievement, relevance, practice, communication, ICT and feedback. Achievement orientation is identified as the most dominant values in these students' mathematics learning. ICT was valued least (relatively) for all three regions.

It should be noticed that although some of the same values have been reported in different countries, their meanings may differ subtly. For example, even though many students valued a 'fun' environment, in Chin and Lin's (2000) study, the term 'fun' was used to describe interesting mathematical problems that elicited Taiwan students' curiosity. But in Hong Kong students valued games and quizzes as a means of maintaining a lively and enjoyable (fun) classroom environment (Law et al. 2012), which is some what different to the use in Taiwan.

Values Taught in Chinese Mathematics Classroom

In the Chinese Mainland, after the Communist party took control in 1949, the educational system from then on was very much influenced by the Soviet Union. This meant that the region was untouched by the Modern Mathematics movement, while basic skills, as well as traditional topics like Euclidean geometry, were emphasized. In the early 2000s education became available to the general mass of the population. China's open door economic policy, first implemented in the mid-1980s, saw an increasing flow of educational ideas from elsewhere, including from Western countries.

Reform-oriented teaching practices, common in the West, require constructivist and inquiry-based classrooms (Ministry of Education of the People's Republic of China 2001, 2003). As well a shift from *the product* (content) emphasis in traditional the Chinese Mainland mathematics teaching, to *process* (ability) has been advocated and is gradually occurring (Wong et al. 2004). Consequently, classroom teaching and learning environments have been changing to meet these new ideas and challenges.

With the mathematics curriculum reforms implemented at the turn of the millennium, and the government's recent proposal for a competency-based curriculum (Zhang and Lam 2017), the Chinese Mainland mathematics classrooms are experiencing considerable changes (Lam et al. 2015). For instance, the use of real-life scenarios, mathematical games and activities, and project-based learning have been advocated by many teacher-educators.

As an extension of the original WIFI study, this study aimed at investigating what kinds of mathematics learning values the Chinese Mainland primary and secondary students now hold, following the curriculum reforms. The findings may afford greater insights into the influence of social-cultural factors on classroom teaching and learning.

Table: Grade and gender of participants

Grades	Gender		N	Percentage (%)
	Girls	Boys		
5–6 (Primary)	320	370	690	28.2
7–8 (Junior secondary)	165	150	315	12.9
10–11 (Senior secondary)	717	726	1,443	58.9
Total	1,202	1,246	2,448*	100.0

*Gender not reported by 70 of the 2,518 students

Research Design and Methodology

A total of 2,518 students from four big cities in the Chinese Mainland were asked to indicate which aspects of mathematics learning they considered most important. The participants were sampled from the cities of Wuhan, Quanzhou, Hangzhou and Dalian, in central, southern, eastern and northern China respectively. A total of 10 government schools were involved in the study; 4 primary, 3 junior secondary and 3 senior secondary. All the students used the same publisher's textbooks that follow the same mathematical curriculum syllabus (MOE 2012). This chapter does not aim at discussing the differences between regions/schools/classrooms. Thus we grouped them together as a whole sample and categorized them only by grades and gender (see Table).

Data for the study were collected using the validated WIFI questionnaire (Seah 2013). A translation and back translation method together with factor analysis was used for metric equivalence checks in the Chinese context (Seah et al. 2017; Zhang et al. 2016). The questionnaire consists of four sections: a 5-point Likert scale consisting of 64 items (section A); 10 items with continuous dimensions (section B); an open-ended scenario-stimulated section (section C); and questions eliciting the students' demographic and personal information (section D). In an earlier analysis (Zhang et al. 2016), a principal component analysis was conducted for section A. In this chapter we report findings from the Chinese Mainland students for section B and section C.

In section B, the semantic differential method was used to measure connotative meaning. Scores were given on a horizontal line with five positions (1, 2, 3, 4 and 5) from left to right. An example is: 'Leaving it to ability when doing mathematics' (on the left) versus 'Putting in effort when doing mathematics' (right). Each side (left or right) represented one value dimension (a mathematical value or a mathematics education value). An independent sample t-test and multivariate analysis of variance were used to analyze the statistical differences between the responses given by students of different genders and grades.

Section C was contextualized using specific scenarios (e.g. Imagine that there is a magic pill. Anyone who takes this pill will become very good at mathematics.) Students are required to nominate what to each of them would be the three most important values in such a situation. The rationale behind this is to allow for the open-ended nature of the responses to provide students with more choice to express their thoughts. The students' responses were coded by using the coding guide in the previous WIFI study and analyzed in terms of the frequency. Three experts (one professor in mathematics education, two experienced mathematics teachers) coded the data individually. We obtained high reliability with a Kappa correlation coefficient of 0.90.

Results

Results in Table indicate that the students tended to place greater value on the process of obtaining the answer to a problem, than on finding the answer itself. They emphasized enjoyment and ability over hard work and effort during learning. Using mathematical concepts to solve a problem was believed to be more important than using a formula to find the answer. Mathematical facts and theories were considered more important than the ideas and practices used in everyday life. The students also believed that remembering mathematical concepts, rules or formulas is more important than creating them. Although learning mathematics from others and exploring mathematics by oneself were considered equally important, the students felt that it is more important for someone (such as the teacher) to provide concrete mathematical examples rather than simply stating the answer. In mathematics teaching, keeping mathematics magical or mystical was valued over merely demonstrating and explaining the subject. The students also felt that mathematics should be used for development or progress rather than simply to explain events.

We found statistically significant differences between the grades for most items in section B, especially between primary and secondary. The primary students tended to value effort, progress, and exploration more highly than the secondary students. Statistically significant differences were also found between the juniors and seniors for half of the 10 items.

Across the population, significant gender differences were found in the responses to six items (Table). The girls valued exploration more than the boys. The boys placed a higher value on fun, ability, rational understanding, and creativity. Further comparison of the three school grades revealed significant gender differences for two items (Q70 and Q71) in the primary group, one item (Q73) in the junior secondary group, and five items (Q68, Q69, Q72 and Q73) in the senior secondary group.

In section C, the students were asked to propose ingredients for a hypothetical magic pill capable of making its taker good at mathematics. Unlike the limited information in section B, section C provided students more choice to express their own thoughts. Of the top five elements cited (see Table), effort was valued most highly across the three groups, with ability and wisdom valued only by the secondary students. Both the primary and junior secondary students valued the use of formulas, whereas only the senior students valued thinking. Interestingly, the boys and girls in each group reported the same top five key elements, but in a different order (see Table).

Table: Differences by grades in students' responses to section B items

Items (score 1–5)	Total		Primary (Grade 5–6)		Junior (Grade 7–8)		Senior (Grade 10–11)		F-test	η^2
	Mean	SD	Mean	SD	Mean	SD	Mean	SD		
66. Process versus product	1.79	1.01	1.74	1.01	1.98	1.01	1.78	1.01	7.00***	0.006
67. Fun versus effort	2.50	1.28	2.39	1.25	2.46	1.23	2.56	1.31	4.18*	0.003
68. Ability versus effort	2.95	1.24	3.41	1.26	3.04	1.21	2.72	1.18	79.08***	0.061
69. Objectism versus rationalism	2.81	1.07	2.75	1.10	2.76	1.00	2.85	1.07	2.67	0.002
70. Facts and theories versus ideas and practice	2.47	1.05	2.56	1.08	2.52	1.03	2.41	1.04	5.04**	0.004
71. Exposition versus exploration	3.02	1.21	3.39	1.20	2.99	1.20	2.86	1.18	48.71***	0.038
72. Recalling versus creating	2.47	1.14	2.49	1.14	2.32	1.04	2.50	1.15	3.35*	0.003
73. Exposition versus exploration	3.72	1.19	3.92	1.19	3.65	1.17	3.64	1.19	14.16***	0.011
74. Openness versus mystery	2.40	1.14	2.50	1.20	2.44	1.07	2.35	1.13	4.51*	0.004
75. Control versus progress	3.16	1.11	3.47	1.14	3.22	1.12	3.01	1.07	42.75***	0.034

Note ***$p < 0.001$, **$p < 0.01$, *$p < 0.05$

Table: Gender differences in students' responses to section B items

Items	Primary					Junior					Senior				
	Girls		Boys		t-test	Girls		Boys		t-test	Girls		Boys		t-test
	Mean	SD	Mean	SD		Mean	SD	Mean	SD		Mean	SD	Mean	SD	
66	1.71	0.99	1.75	1.03	−0.59	1.89	0.92	2.07	1.10	−1.52	1.79	0.97	1.73	1.04	1.20
67	2.38	1.20	2.40	1.28	−0.24	2.54	1.21	2.33	1.25	1.47	2.63	1.31	2.50	1.30	1.89
68	3.46	1.23	3.38	1.26	0.78	3.13	1.15	2.95	1.28	1.31	2.79	1.12	2.65	1.24	**2.35**
69	2.83	1.00	2.68	1.17	1.84	2.80	0.97	2.71	1.06	0.84	3.03	1.02	2.69	1.11	**6.06*** **
70	2.67	1.04	2.46	1.12	**2.56***	2.55	0.97	2.50	1.10	0.47	2.43	1.01	2.40	1.07	0.41
71	3.50	1.14	3.29	1.25	**2.32***	3.12	1.17	2.86	1.22	1.93	2.87	1.13	2.85	1.24	0.28
72	2.42	1.12	2.57	1.15	−1.70	2.23	0.96	2.43	1.13	−1.70	2.39	1.07	2.60	1.22	**−3.32*** **
73	4.02	1.13	3.85	1.24	1.87	3.79	1.05	3.51	1.28	**2.11***	3.75	1.09	3.56	1.26	**3.20****
74	2.49	1.21	2.50	1.18	−0.11	2.43	1.10	2.45	1.03	−0.14	2.32	1.09	2.37	1.18	−0.77
75	3.49	1.08	3.44	1.19	0.60	3.30	1.00	3.18	1.22	0.95	3.06	0.98	2.97	1.14	1.72

Note ***$p < 0.001$, **$p < 0.01$, *$p < 0.05$

Table: Frequency (%) comparisons between groups on key elements in section C

Overall top 5	Key element	Primary	Key element	Junior secondary	Key element	Senior secondary
1st	Effort	133(30.8%)	Effort	62(17.5%)	Ability	405(29.2%)
2nd	Diligence	75(17.4%)	Ability	44(12.4%)	Thinking	323(23.3%)
3rd	Formula	69(16.0%)	Formula	42(11.8%)	Effort	261(18.8%)
4th	Smartness	68(15.7%)	Wisdom	41(11.5%)	Wisdom	219(15.8%)
5th	Memory	59(13.7%)	Knowledge	41(11.5%)	Memory	199(14.4%)

Table: Gender difference between groups with respect to the key elements in section C

Overall top 5	Primary		Junior secondary		Senior secondary	
	Girls	Boys	Girls	Boys	Girls	Boys
Ability	5(2.60%)	21(9.01%)	18(13.4%)	24(19.2%)	258(34.9%)	143(22.7%)
Effort	72(37.5%)	24(10.3%)	32(23.9%)	27(21.6%)	149(20.2%)	111(17.6%)
Thinking	12(6.25%)	19(8.15%)	10(7.46%)	6(4.8%)	187(25.3%)	131(20.8%)
Wisdom	32(16.7%)	22(9.44%)	21(15.7%)	17(13.6%)	130(17.6%)	85(13.5%)
Memory	24(12.5%)	35(15.0%)	14(10.4%)	3(2.4%)	117(15.8%)	30(4.77%)

The findings from section C showed many consistencies with the findings for section B. For example, efforts and memory are both valued by most students in sections B and C. An interesting finding is that the value of exposition was given more emphasis in section B, while the element of teacher(s) is seldom mentioned by students in section C. Smartness or wisdom entered one of top five important elements in section C, this natural ability seems more valued by primary students in section B compared to secondary students.

Discussion and Conclusion

Understanding more about values is key to generating possibilities for mathematics teaching. These findings revealed that both primary and secondary Chinese Mainland students tended to value process, pleasure, ability, facts and theories, recall, and exploration over their respective opposing dimensions, product, effort, ideas, create and exposition respectively. The value dimensions of process, facts and theories, and recall are very similar to the findings in the previous study in section A (Zhang et al. 2016). For example, in section A, students value achievement, which indicated they thought "knowing the steps of a solution", "knowing which formula to use" and "memorizing the facts" are important for their mathematics learning. These can also be reflected in students' choices in Q70 and Q72 of section B where students tended to value truths and facts in mathematics and remembering mathematical ideas, concepts and rules. In section C, we could also see these with formula and memory in the top five choices of students' values. Interestingly Zhang et al.

(2016) reported that both the relevance to mathematics (e.g., stories about mathematics, hands-on activities, and mathematics puzzles) and practice were important for students learning mathematics. But here in Q67 of section B, when they needed to make a choice on fun and effort when doing mathematics, students tended to value feeling relax or having fun more than hard work. The emphasis on pleasure, process, and exploration also echoed the findings reported by Zhang (2014). The students believed that an enjoyable atmosphere is important to mathematics learning and felt that students need to be involved in classroom activities. These priorities were echoed in the Chinese Mainland recent curriculum reforms (MOE 2001, 2003, 2012). Of the mathematical value dimensions tested (e.g., in Q69, Q74, and Q75), objectism (be more pragmatic and concrete), openness (sharing mathematics with others), and progress (using mathematics for development) were highlighted.

This finding, together with the students' emphasis on facts and theories and the process of recall, resembles the results reported in previous studies (Seah and Peng 2012; Zhang 2014). There the students believed that an effective mathematics teacher must present, demonstrate, and explain mathematics-related information clearly. So not only gaining mathematical knowledge, but also learning how to learn was considered important. Although constructivism and student-centered learning are promoted in curriculum reforms in the Chinese Mainland, these results suggest that students still prefer the teacher-led approach. The product–process dichotomy has generally been regarded as the major distinction between Eastern and Western mathematics teaching (Leung 2001), with Chinese mathematics classroom being teacher-oriented/teacher-centered. However, the notion of teacher-led, yet with a student learning focus, may be a more suitable description of the Chinese Mainland lessons script (see Wake and Pampaka 2008; Wong 2009).

In our former study (Zhang et al. 2016), only primary students' views are investigated. However it appears that as they progress through school their value choices do change. In section B, statistically significant differences were found between both the primary and secondary students and the junior secondary and senior secondary students. These differences may be explained by the students' own growth and development as they move on from primary to secondary and experience changes in their learning environment, teaching approaches, and cultural context.

Just how the students' environment changes as they progress might be worth further exploration. In primary classrooms students are told to work harder and be diligent more frequently than secondary classrooms. As well in secondary years ability becomes more important to students. These are hints of the changes that are present but further work is needed.

It seems that what students value in their mathematics learning is probably influenced by how they have been taught in the classroom, as well as their existing conceptions of mathematics (Wong et al. 2002, 2016; Zhang and Wong 2015). If a student sees mathematics as a set of rules, facts, and procedures, his or her learning approach and understanding probably will be more instrumental, and will probably also value this approach to learning mathematics. In contrast, students in a constructivist classroom may hold a broader conception of mathematics and show more profound learning motives and strategies. Some results suggest that such students have more positive learning attitudes, place greater value on mathematics, and are less anxious about mathematics learning (Ding and Wong 2012). Clearly these ideas are laced through with values, and need to be more thoroughly explored in the Chinese Mainland context.

The Chinese society is widely believed to hold education in high regard, to value effort, and to be achievement oriented (Leung 2001). This was supported by the results from section C. Chinese students have an examination-oriented mentality, influenced by Chinese cultural values such as a social achievement orientation and an emphasis on diligence, effort, and collectivism. Interestingly, although teaching practices and parents' expectations have been shown to affect children's learning (Hauser-Cram et al. 2003; Rosenthal and Jacobson 1968; Rubie-Davies et al. 2010), the students in this study did not explicitly indicate the importance of these external factors for their mathematics learning. Specifically, the top five elements listed in response to section C included neither teachers nor parents. Instead, the students tended to focus on themselves: e.g., their own ability, knowledge, memory and wisdom. It may be these students believed that internal factors are more important to learning than external factors, and the help and support from their teachers or parents maybe perceived by them as not being as beneficial as it appears to be elsewhere. Hence this suggests that a further investigation should focus on these external factors.

In this study we focused on the little researched area of student values. However how students learn values has not been addressed and awaits more insightful study. Does value negotiation occur between teachers and students in the teaching and learning of mathematics in the Chinese Mainland? How and in what contexts can teachers foster such negotiation? By exploring what values students do seem to exhibit at various school levels, as was done here, may give a good guide in exploring these further questions. The answers to all these questions will deepen our understanding of students' mathematics learning environments, both in the Chinese Mainland and more widely, and help teachers to design their teaching more effectively.

Acknowledgements The research work is partially supported by the Sumitomo Foundation Grant-168066 and the Internal Research Grant (RG93/2017-2018R) of the Education University of Hong Kong. Special thanks are due to the editors and anonymous reviewers and for their helpful and constructive comments.

References

Atweh, B., & Seah, W. T. (2008). *Theorizing values and their study in mathematics education.* Paper presented at the AARE Conference, Fremantle, Australia.

Bishop, A. J. (1988). *Mathematical enculturation: A cultural perspective on mathematics education.* Dordrecht: Kluwer.

Bishop, A. J. (1999). Mathematics teaching and values education: An intersection in need of research. *ZDM Mathematics Education, 31*(1), 1–4.

Bishop, A. J., Clarkson, P., FitzSimons, G., & Seah, W. T. (2001). Studying values in mathematics education: Aspects of the VAMP project. In J. Novotna (Ed.), *European research in mathematics education* (Vol. 2, pp. 368–376). Prague: Charles University.

Chin, C., & Lin, F. L. (2000). *Values and values statement emerged in students' preferences on test items: A case study from mathematical induction.* Paper presented at the history and pedagogy of mathematics 2000 conference, Taipei, Taiwan.

Dede, Y. (2006). Values in Turkish middle school mathematics textbooks. *Quality & Quantity, 40,* 331–359.

Dede, Y. (2015). Comparing primary and secondary mathematics teachers' preferences regarding values about mathematics teaching in Turkey and Germany. *International Journal of Science and Mathematics Education., 13*(1), 227–255.

Ding, R., & Wong, N. Y. (2012). The learning environment in the Chinese mathematics classroom. In Y. Li & R. Huang (Eds.), *How Chinese teach mathematics and improve teaching* (pp. 150–164). New York: Routledge.

Hauser-Cram, P., Sirin, S. R., & Stipek, D. (2003). When teachers' and parents' values differ: Teachers' ratings of academic competence in children from low-income families. *Journal of Educational Psychology, 95,* 813–820.

Lam, C. C., Wong, N. Y., Ding, R., Li, S. P. T., & Ma, Y. (2015). Basic education mathematics curriculum reform in the greater Chinese region — Trends and lessons learned. In B. Sriraman, J. Cai, K-H. Lee, L. Fan, Y. Shimuzu, C. S. Lim, & K. Subramanium (Eds.), *The first sourcebook on Asian research in mathematics education: China, Korea, Singapore, Japan, Malaysia and India* (pp. 305–335). Charlotte, New Carolina, U.S.A: Information Age Publishing.

Law, H. Y., Wong, N. Y., & Lee, L. N. Y. (2012). A study into espoused values in Hong Kong mathematics classrooms. *ZDM Mathematics Education, 44*(1), 45–57.

Leung, F. K. S. (2001). In search of an East Asian identity in mathematics education. *Educational Studies in Mathematics, 47,* 35–51.

Lim, C. S. (2015). Riding the third wave: Negotiating teacher and students' value preferences relating to effective mathematics lesson. In S. J. Cho (Ed.), *Selected regular lectures: 12th international congress on mathematical education* (pp. 471–485). Cham: Springer.

Ministry of Education of the People's Republic of China. (2001). *Mathematics curriculum standard for compulsory education stage (experimental version)* [in Chinese]. Beijing, People's Republic of China: Beijing Normal University Press.

Ministry of Education of the People's Republic of China. (2003). *Mathematics curriculum standard for normal senior high schools* [in Chinese]. Beijing, People's Republic of China: People's Education Press.

Ministry of Education of the People's Republic of China. (2012). *Mathematics curriculum standard for compulsory education stage* (2011 ed.) [in Chinese]. Beijing, People's Republic of China: Beijing Normal University Press.

Opdenakker, M. C., & van Damme, J. (2006). Teacher characteristics and teaching styles as effectiveness enhancing factors of classroom practice. *Teaching and Teacher Education, 22,* 1–21.

Rosenthal, R. J., & Jacobson, L. (1968). *Pygmalion in the classroom: Teacher expectations and pupils' intellectual development.* New York: Holt, Rinehart & Winston.

Rubie-Davies, C. M., Peterson, E., Irving, E., Widdowson, D., & Dixon, R. (2010). Expectations of achievement: Student, teacher and parent perceptions. *Research in Education, 83*(1), 36–53.

Seah, W. T. (2013). Assessing values in mathematics education. In A. M. Lindmeier & A. Heinze (Eds.), *Proceedings of 37th conference of the international group for the psychology of mathematics education* (Vol. 4, pp. 193–200). Kiel, Germany: PME.

Seah, W. T., & Andersson, A. (2015). Valuing diversity in mathematics pedagogy through the volitional nature and alignment of values. In A. Bishop, T. Barkatsas, & H. Tan (Eds.), *Diversity in mathematics education: Towards inclusive practices* (pp. 167–183). New York, NY: Springer.

Seah, W. T., Andersson, A., Bishop, A., & Clarkson, P. (2016). What would the mathematics curriculum look like if values were the focus? *For the Learning of Mathematics, 36*(1), 14–20.

Seah, W. T., Baba, T., & Zhang, Q. P. (2017). The WIFI study: Students' valuing of mathematics learning in Hong Kong and Japan. In J. W. Son, T. Watanabe, & J. J. Lo (Eds.), *What matters? Research trends in international comparative studies in mathematics education* (pp. 333–354). Cham, Switzerland: Springer.

Seah, W. T., & Peng, A. (2012). What students outside Asia value in effective mathematics lessons: A scoping study. *ZDM, 44*(1), 71–82.

Seah, W. T., & Wong, N. Y. (2012). What students value in effective mathematics learning: A 'Third Wave Project' research study. *ZDM Mathematics Education, 44*(1), 33–43.

Shinno, Y., Kinone, C., & Baba, T. (2014). Exploring 'what Japanese students find important in mathematics learning' based on the Third Wave Project. In C. Nicol, S. Oesterle, P. Liljedahl, & D. Allan (Eds.), *Proceedings of the joint meeting of PME 38 and PME-NA 36* (Vol. 5, pp. 169–176). Vancouver, Canada: PME.

Wake, G., & Pampaka, M. (2008). The central role of the teacher - even in student centered pedagogies. In O. Figueras, J. Cortina, S. Alatorre, T. Rojano, & A. Sepu´lveda (Eds.), *Proceedings of the Joint Meeting of PME 32/PME-NA 30* (pp. 377–384). Mexico: PME.

Wong, N. Y. (2009). Exemplary mathematics lessons: What lessons we can learn from them? *ZDM Mathematics Education, 41,* 379–384.

Wong, N. Y., Ding, R., & Zhang, Q. P. (2016). From Classroom environment to conception of mathematics. In R. B. King & A. B. I. Bernardo (Eds.), *The psychology of Asian learners: A festschrift in honor of David Watkins* (pp. 541–557). Singapore: Springer.

Wong, N. Y., Han, J. W., & Lee, P. Y. (2004). The mathematics curriculum: Towards globalisation or Westernisation? In L. Fan, N. Y. Wong, J. Cai, & S. Li (Eds.), *How Chinese learn mathematics: Perspectives from insiders* (pp. 27–70). Singapore: World Scientific.

Wong, N. Y., Marton, F., Wong, K. M., & Lam, C. C. (2002). The lived space of mathematics learning. *Journal of Mathematical Behavior, 21,* 25–47.

Zhang, Q., Barkatsas, T., Law, H. Y., Leu, Y.-C., Seah, W. T., & Wong, N. Y. (2016). What primary students in the Chinese Mainland, Hong Kong and Taiwan value in mathematics learning: A comparative analysis. *International Journal of Science and Mathematics Education., 14*(5), 907–924.

Zhang, Q. P. (2014). The third wave: Chinese students' values in effective mathematics teaching in two secondary schools. *Journal of the Korea Society of Mathematical Education Series D: Research in Mathematical Education, 18*(3), 209–221.

Zhang, Q. P., & Lam, C. C. (2017). Reflections on competency-based mathematics education (in Chinese). *People's Education, 6,* 43–47.

Zhang, Q. P., & Wong, N. Y. (2015). Beliefs about mathematics, mathematics knowledge and approaches to teaching among Chinese teachers. In L. Fan, N. Y. Wong, J. Cai., & S. Li (Eds.), *How Chinese teach mathematics: Perspectives from insiders* (pp. 457–492). Singapore: World Scientific.

5

Methodological Issues in the Investigation of Values in Mathematics

Yip-Cheung Chan and Ngai-Ying Wong

Abstract In this chapter, we first provide an overview on our previous investigations, then the methodologies employed are evaluated. A variety of research methodologies such as questionnaires, open-ended questions, interviews, classroom observation, episode writing, and mind maps have been employed at different stages of our intellectual journey. These methods are effective tools and helped us understand students' and teachers' beliefs and values about mathematics, mathematics learning and teaching. However, one's value is hidden and therefore not apparent. It is a deep-down notion in the sense that it is not easily unfolded. Thus, if we want to have a more holistic understanding on teachers' and students' values (in mathematics), we need a sharper methodology in which the participant is forced to make decision under a dilemmatic situation so that their values can be revealed. After providing examples of such situation that we used, we would argue that the use of hypothetical situations can enhance the investigations of values and related issues. What we used include (a) asking whether one is doing mathematics when provided with some hypothetical situations, (b) what a teacher would respond if his/her students asked whether one is doing mathematics in the given hypothetical situations, (c) how one respond to some famous mathematicians' quotes in which some controversial situations are involved, (d) hypothetical classroom situations and (e) hypothetical lesson planning. We believe that our discussions can enrich the study of values in mathematics education.

Keywords Research methodologies · Beliefs about mathematics · Values in mathematics · Hypothetical situations

---

Y.-C. Chan (✉)
Department of Curriculum and Instruction, The Chinese University of Hong Kong, Shatin, Hong Kong
e-mail: mathchan@cuhk.edu.hk

N.-Y. Wong
The Education University of Hong Kong, Tai Po, Hong Kong

Theoretical Premises

Despite the fact that beliefs and values are important notions in mathematics education research, the methodologies for investigating these notions are still developing (Leder et al. 2002). By reviewing existing literatures related to this area, we found that beliefs and values are closely related but values are deeper than beliefs. According to Philipp (2007), value is "a belief one holds deeply, even to the point of cherishing, and acts upon" (p. 259). On the other hand, we think that one's value is hidden and therefore not apparent. Thus, it is not easy to be unfolded by simple methods such as observing one's behaviour or completing a questionnaire. In order to have a holistic understanding on students' and teachers' values in mathematics, a variety of research methodologies are needed. This is precisely the theme of this chapter. We will first review and reflect on the methodologies used in our own related studies. Then, we will argue that hypothetical situations can be a methodology which is complementary to other frequently used methodologies such as questionnaires and interviews. By hypothetical, we mean that the researcher confronts the participant (usually by means of interview) with a situation in which the participant may not actually have a chance to encounter before. More importantly, the situation is usually put in an extreme (dilemmatic) situation in which the participant needs to make a choice among different options. We believe that such choice can provide a 'window' for unfolding one's values.

Beliefs and Values as Affective Subdomains in Mathematics

Affect in general—as well as beliefs and values in particular—have drawn attention to the mathematics education community in recent decades. While Daskalogianni and Simpson (2000) exclaimed on the non-consensus definition of attitude that "… almost reducing it [the concept of attitude] to the pseudo-definition 'attitude is what attitude questionnaires measure'" (p. 217), the same might be true for affects, beliefs and values (see for instance, Hannula 2012). The interrelationships among these notions began to take shape in McLeod's seminal work (1992). Beliefs, attitudes and emotions are identified as three affective subdomains among which these three subdomains increased in the levels of affective involvement and intensity of response whereas they decreased in the levels of cognitive involvement and response stability. Beliefs play a significant role in students' development of attitudinal and emotional responses to mathematics. On the other hand, automatized repeated emotions can become an attitude.

Under the above framework, Goldin (2000) proposed the notion of affective pathways which sharpen the interrelationship among different affective subdomains.

It emphasizes the interaction of affective states and cognitive representations during mathematical problem solving. DeBellis and Goldin (2006) further developed a tetrahedral model which can be regarded as an extension of McLeod's framework. This model highlights the interacting relationship of **values** with the other three affective subdomains. According to DeBellis and Goldin (2006), "*values*, including *ethics* and *morals*, refer to the deep, 'personal truths' or commitments cherished by individuals" (p. 135, italic in its original). In around that period of time, Hannula (2012) re-conceptualized the notion of attitude, in which emotions, expectations and values are included as different aspects for analyzing one's attitude and its changes.

The Close Relationship Between Beliefs and Values

While DeBellis, Goldin and Hannula identified beliefs and values as subdomains of affect in mathematics, the close relationship between beliefs and values are emphasized by different scholars (see for instance, Barkatsas et al. 2018). Beliefs are usually related with one's knowledge or conceptions about something (Furinghetti 1998) whereas values are something that we consider to be important and worthwhile (Bardi and Schwartz 2003). This distinction and their relationships are stated by Philipp (2007). Beliefs are considered as "psychologically held understandings, premises, or propositions about the world that are thought to be true". In contrast, values are "the worth of something [and] a belief one holds deeply, even to the point of cherishing, and acts upon" (p. 259). It highlights the fact that values are **deeper** than beliefs because those beliefs held deeply by an individual are identified as one's values.

Even as early as 1990s, Bishop (1999) has pointed out that "values in mathematics education are the **deep** affective qualities which education fosters through the school subject of mathematics" (p. 2, emphasis added). Furthermore, values are something that are internalized as important and worthwhile, and thus they define how one sees the world (Seah and Andersson 2015).

Worldview is another notion commonly found in literature, which has close resemblance with value. It is "one's comprehensive set of beliefs about the nature of reality and how one should live in the light of those beliefs" (Heie 2002, p. 99). In other words, a person's worldview is an amalgamated product of multiple sources, including but not limited to one's beliefs and values. In the context of mathematics teaching and learning, one should not only focus on values in the subject discipline of mathematics but also need to address the values in general such as educational values, sociocultural values, or even religious values. This idea echoes to Bishop et al. (1999)'s identification of three kinds of values in mathematics teaching, that is: the general educational, the mathematical, and the specifically mathematics educational.

Although the definitions of value may vary among different scholars, all these definitions suggest that value is so deep down that it is not easily unfolded.

Our Journey on Investigating Beliefs and Values in Mathematics Education

Although we recognize that different scholars each has his/her intellectual journeys, we believe that our own journey is enough to reflect the methodological challenges one is facing when researching on beliefs and values. Thus, instead of doing a comprehensive literature review (which is not the goal of this chapter), we will review our journey on investigating mathematics-related affects (especially beliefs and values) in which the authors of this chapter have been involved. We will first introduce an overarching framework which guides our studies and then review the methodologies which have been employed.

The Lived Space of Mathematics Learning

A group of researchers in Hong Kong, led by the second author of this chapter, launched a project on mathematics beliefs/values in the mid-1990s. To begin with, they investigated students' conceptions of mathematics and their beliefs about mathematics/mathematics learning. How such beliefs affect their problem solving behaviour was also investigated. Teachers' beliefs were investigated at the same time, which later extended to teachers' teaching behaviours as well as their knowledge (Subject Knowledge and Pedagogical Content Knowledge included). How teachers' religion and their worldview might influence their teaching philosophy (including beliefs and values) as well as teaching per se was the most recent endeavour of the two authors of this chapter. All these investigations can be framed under the notion of *lived space* of mathematics learning (Wong et al. 2002).

As depicted in Fig., the series of studies was nicely conceptualized into the framework of the *lived space*. First, how a phenomenon (learning phenomenon included) is perceived by either an individual or by a group generates a *space* of understanding or conception of that particular phenomenon. In terms of learning, that *space* would constitute an *outcome space* regarding that phenomenon (Marton and Booth 1997). The antecedent of this *outcome space* is students' *lived space* (which is largely shaped by the teacher) which is the result of their learning experiences. By reinterpreting earlier findings using the theoretical lens offered by phenomenography, we can conclude that a broader *lived space* leads to a richer *outcome space* of student learning (Marton and Booth 1997).

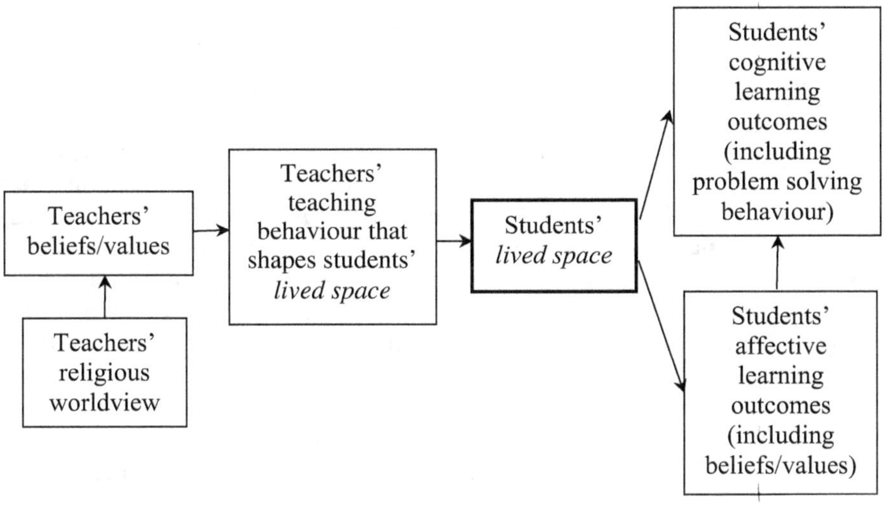

Fig. The *lived space* of mathematics learning

Methodologies Used in Our Studies

Different methodologies have been used at different stages of our studies. Open-ended questions such as "Mathematics is …" was used in our study on students' beliefs about mathematics (Wong 1993). In our next study on students' beliefs about understanding mathematics, both open-ended questions (for instance, "When will you consider having understood a certain mathematics concept?") and episode writing (for instance, recall and write down any instance in which the participants understood a mathematics topic, formula, rule or problem) were employed (Wong and Watkins 2001). Besides arriving at other results, *the Mathematics Classroom Environment Scale* was developed. At that time, **we realized that it is not easy for students to tell us what mathematics is**. We adapted the ideas of Kouba and McDonald (1991) by utilizing hypothetical situations to investigate students' conception of mathematics. An example of such situations is "An elder sister lifted her younger brother. She said that he must weigh about 30 lb less than she. Did she do mathematics?" (Lam et al. 1999). Another inventory *The Conception of Mathematics Scale* was developed which was used in a number of studies, including an Ed.D. thesis on the use of history in mathematics teaching (Cheung 2014). In addition to the inventory, mind maps were used to tap students' conceptual change.

The above studies focused solely on students' beliefs about mathematics (and about mathematics learning). We moved one step forward to explore the linkage between beliefs and their mathematics learning outcomes, in particular, their performances in mathematics problems (Wong et al. 2002). In addition to computational and word problems, open mathematics problems were used (Cai 1995). Clinical interviews were conducted after the problem-solving process. Students' confined conception of mathematics would lead them to approach mathematics problems in a mechanistically way, solving them by picking out routines. We hypothesized that this is the consequence of the narrow *lived space* and thus we proceeded to investigate the conception of mathematics/mathematics learning and teaching among the teachers.

We moved from studying **students** to studying **teachers**. Since teachers are adults and are trained in mathematics/mathematics education, **in spite the possibility that they just offer 'model answers' in answering what is mathematics, (projected) hypothetical situations were once again employed**, that is, asking teachers their expectation of their students' reactions when they were confronted with the above hypothetical situations and what would be their own responses. In addition, the teachers were asked to comment on quotes from famous mathematicians (which are often controversial). An example of such a quote is "The moving power of mathematical invention is not reasoning but imagination" (A. DeMorgan, cited in Graves 1889, p. 219). The study found that the teachers hold a slightly broader conception of mathematics than the students, yet there is much concord between the two (Wong et al. 2003). Semi-structured interviews (about mathematics, learning and teaching) were used again in Cai et al. (2009).

The close resemblance between the students' conception of mathematics and the teachers' conception of mathematics reinforced our premise that teachers' conception of mathematics contributes to the students' conception of mathematics. We need a further step to trace **how** the *lived space* of the students was shaped by the teachers. To achieve this, we entered the classroom. This is precisely the theme of Q. T. Wong's M.Phil. thesis (2003) (a summary of her work can be found in Wong et al. 2009). In her project, different methods were used to address different research agendas. In particular, the projected hypothetical situations of whether 'one is doing mathematics' (transformed to asking the teachers) were used once again. She also conducted classroom observation and follow-up interviews. In addition, she used hypothetical situations in classroom teaching to solicit participants' beliefs about mathematics teaching. First, she collected questions which often arouse debates among mathematics teachers. Then, she used these issues to stimulate her participants by asking them how they would react. The following is an example of the interview questions.

> When you are marking students' homework like "Each apple costs \$3. What is the total price of 2 apples?", which of the following(s) – \$3 × 2, \$ (3 × 2) or \$ (2 × 3) – do you regard as incorrect and would marks be deducted? And why? (Wong 2003, p.120)

Zhang's Ph.D. study (2010) moved one step further to include teachers' knowledge (a summary of his work can be found in Wong et al. 2009). He found that teachers' beliefs and their knowledge co-contribute to the shaping of students' *lived space*.

The Third Wave Project, which was initiated by Wee Tiong Seah, aimed at cross-regional comparisons of students' and teachers' values in mathematics learning and their co-valuing (Seah and Wong 2012). In this project, not only the notion of beliefs was extended to values, this is the first study (among those that we are involved with) that both students and teachers were investigated in a single study. In addition to widely used methods such as teacher journals, lesson observations and semi-structured interviews (see for instance, Wiegerová 2013), a new method involving 'photo-voice' (Lim 2010) was utilized. Students were asked to take photo snapshots in 'aha' moments, that is moments that the students felt inspiring (including suddenly understood something) during the lesson. These photo snapshots formed the basis of both teacher and student post-lesson interviews. This kind of stimulus-recalled interviews enabled us to capture students' and teachers' espoused values in effective mathematics classrooms (Law et al. 2012).

This Project drew our attention to the relationships between values in mathematics and one's worldviews, teachers' religious beliefs in particular. Questionnaire and semi-structured interview were used in the first two studies (Chan and Wong 2014; Leu et al. 2015). Both studies led us to conduct our main study (Chan and Wong 2016) which investigated teachers' **enacted** values in mathematics that may be related to their religious values. In this study, hypothetical lesson planning was used. The respondents were requested to design a mathematics lesson which referred to their religious beliefs in mathematics teaching. It was hypothetical because the respondents could assume that they were permitted to do whatever things related to their religion in that lesson (which may or may not be true). This hypothetical lesson planning formed the basis of the follow-up semi-structured interviews which intended to capture the teachers' interplay between their religious values and their enacted values in mathematics.

While one can refer to our previous publications for details of examples of these hypothetical situations, let us provide another example of using hypothetical situations. First, participants were provided by a hypothetical mathematics lesson, in the form of a script (lesson plan), which is pre-analyzed and truncated into routines. At the junction of different routines, the participants were confronted with hypothetical questions like "If you were that teacher, what would you do next, why?"; "If you are not allowed to do what you proposed to do, what other teaching strategies can you think of?"; "If at this point, a student reflected that s/he does not understand, what other ways you can explain again?", "Please offer as many alternatives as possible and compare the strengths and weaknesses of these strategies". Since the entire situation is hypothetical, it leaves more room for both designing the lesson plan and the setting of the questions (Cai and Wong 2012).

Methodology Revisited

Though the description above is confined to the studies conducted by the authors of this chapter, it could project a picture of what other researchers are using because most of the studies were collaborative with other researchers overseas. In fact, if one explores standard texts on research methods like Punch (1998), the methodologies listed do not go beyond what we tried, thus ours may be much richer. We attempted many methodologies (sometimes used several methods simultaneously) because we realised that it is not easy to tap into one's values which are deep down and hidden in one's mind. The methods that we have attempted include conventional questionnaires, semi-structured interviews, teachers' journals, open-ended questions, episode writing, snapshots of critical moments, scenario-stimulated responses, clinical interview and hypothetical situations.

Apparently, no single means is more powerful than the others. It is common knowledge that each methodology has its limitations.

As the researcher is (part of) the instrument of qualitative methods, data-sensitivity of the researcher is crucial. Qualitative methods often face the challenge of being subjective. Results are usually not generalizable. Since the labour involved is intensive, the number of respondents is often limited. Social desirability (that is, the tendency of giving a socially acceptable response) is another issue if data collection involves face to face meetings (e.g., interviews). One is referred to standard texts like Creswell (2018), Denzin and Lincoln (2005), Kvale (1996), LeCompte and Preissle (1993), Rubin and Rubin (2012) and Silverman (2001) for easy reference. It may make the issue more intense when the topic is sensitive. Values could be one such issue.

Quantitative methods seem to be more 'objective'. However, quantitative means of data collection were often conducted in one-way paper-and-pencil fashion. Therefore, idiosyncratic issues (e.g. misinterpretation and having different respondents perceiving the same item differently) may arise. These issues could be caused by gender, cultural and other individual differences. We cannot rule out the possibility that some groups of respondents (Chinese in particular, who are typically modest) tend to pick middle response scores (Harzing 2006). Furthermore, quantitative methods such as Likert-type inventories appear to be objective, yet their responses are restricted by the questionnaire framework (construct dimensions) and guided by the questionnaire items. Quantitative methods are not as exploratory as qualitative methods. Thus, it is not easy to discover aspects beyond what is laid down by the questionnaire. The issue of social desirability, though not as prominent, is still there (Bernardi 2006; Fisher and Katz 2000; Lalwani et al. 2006).

One needs multiple sources of data to triangulate, piecing together different facets (from sources of data) into a relatively impartial picture. Therefore, mixed methods are becoming increasingly popular. Thus, we changed methodologies as the research progressed. We chose methodologies based on our specific needs (e.g., the number of participants, their age, as well as the sensitivity of the topic) because no single methodology serves all our needs.

On top of the above issues (including social desirability), what makes value research more challenging is that values may not be easily articulated. We need to go further to unfold such a notion which is deep down in one's mind. We are not saying that the above approaches and methods do not work. Nevertheless, some of these methods are just a setting to solicit informants' ideas. For instance, one can incorporate open-ended questions or episode writings in case studies or in focused group interviews. However, what is clear is that we need to overcome issues like having loaded questions, being too rely on self-reporting and social desirability. As such, we explored different possibilities in our academic journey.

Hypothetical situations are one that might fill the above gap. We are not advocating that the use of hypothetical situations is so powerful to replace all other means.

Rather, hypothetical situations can only be incorporated into other methodologies, e.g., interviews, or even paper-and-pencil 'questionnaires' (such as episode writing).

Let us re-iterate on the various kinds of hypothetical situations we used in our journey. Initially, we employed the method used by Kouba and McDonald (1991). With this method, students were confronted with situations and asked whether they considered them as doing mathematics. We adapted this by asking the teachers their expectations of students' responses when the students were asked whether these situations were considered as doing mathematics. Then we proceeded to use quotations of famous mathematicians in history. Hypothetical situations that concerns frequently debated issues in mathematics departments (within a school) were also used. Whether one regards these statements as mathematically correct caused much controversy in the wider mathematics teacher circle too. We confronted our informants about these frequently asked questions to unfold their beliefs and values about mathematics education. Other types of hypothetical situations were used in subsequent studies on how teachers' values about mathematics education are related to their religious values. Our respondents were asked to develop a hypothetical mathematics lesson that may express their religious values in the lesson.

The gist of hypothetical situations is that we provide participants with a situation that pushes them to an extreme so that they need to review their own values in order to make a response, thus revealing these values from the bottom of their hearts. In contrast, naturalistic situations such as observation of normal classrooms could be too routinized and hence unable to confront the participants through dilemmatic situations. In addition, the respondents' practical considerations such as administration or curriculum requirements would make it more difficult to extract the values behind the scene. We honour other data collection methods, but we do think that the use of hypothetical situations provides a means particularly relevant to value studies.

Conclusion

Values are deep down construct that no single method could possibly portray the whole picture. (Thus, some scholars even use the term 'value system'.) It is imperative in research involving values to triangulate the data with different methods. By triangulation we do not simply mean having different points (data collection methods) to cross check each other but to re-construct the 'real' picture through various points. We do not infer that the use of hypothetical situations is a panacea; we only see its high potential in value research. We would argue that values should only be explored by using such in-depth methodologies.

References

Bardi, A., & Schwartz, S. H. (2003). Values and behavior: Strength and structure of relations. *Personality and Social Psychology Bulletin, 29,* 1207–1220.

Barkatsas, T., Law, H. Y., Wong, N. Y., & Seah, W. T. (2018). Valuing from student's perspectives as a lens to understand mathematics learning: The case of Hong Kong. In B. Rott, G. Törner, J. Peters-Dasdemir, A. Moller, & Safrüdiannur (Eds.), *Views and beliefs in mathematics education: The role of beliefs in the classroom* (pp. 43–53). Cham, Switerland: Springer.

Bernardi, R. A. (2006). Associations between Hofstede's cultural constructs and social desirability response bias. *Journal of Business Ethics, 65*(1), 43–53.

Bishop, A. (1999). Mathematics teaching and values education—An intersection in need of research. *Zentralblatt für Didaktik der Mathematik, 31*(1), 1–4.

Bishop, A., FitzSimons, G., Seah, W. T., & Clarkson, P. (1999). *Values in mathematics education: Making values teaching explicit in the mathematics classroom.* Paper presented at the Combined Annual Meeting of the Australian Association for Research in Education and the New Zealand Association for Research in Education, Melbourne, Australia.

Cai, J. (1995). *A cognitive analysis of U.S. and Chinese students' mathematical performance on tasks involving computation, simple problem solving, and complex problem solving.* Journal for Research in Mathematics Education monograph series 7. Reston, VA: National Council of Teachers of Mathematics.

Cai, J., & Wong, N. Y. (2012). Effective mathematics teaching: Conceptualisation, research, and reflections. In W. Blum, R. B. Ferri, & K. Maaß (Eds.), *Mathematikunterricht im Kontext von Realität, Kultur und Lehrerprofessionalität (Mathematics lesson in the context of reality, culture and teacher professionalism)* (pp. 294–303). Wiesbaden: Springer Spektrum.

Cai, J., Perry, B., Wong, N. Y., & Wang, T. (2009). What is effective teaching? A study of experienced mathematics teachers from Australia, the Mainland China, Hong Kong-China, and the United States. In J. Cai, G. Kaiser, B. Perry, & N. Y. Wong (Eds.), *Effective mathematics teaching from teachers' perspectives: National and cross-national studies* (pp. 1–36). Rotterdam: Sense Publications.

Chan, Y. C., & Wong, N. Y. (2014). Worldviews, religions, and beliefs about teaching and learning: Perception of mathematics teachers with different religious backgrounds. *Educational Studies in Mathematics, 87*(3), 251–277.

Chan, Y. C., & Wong, N. Y. (2016). A multiple case study of teachers referring to their own religious beliefs in mathematics teaching. *International Journal of Children's Spirituality, 21*(3–4), 243–255.

Cheung, W. S. (2014). *The effect on students' mathematical beliefs by integrating history of mathematics in the classroom* [in Chinese]. Unpublished EdD thesis, The Chinese University of Hong Kong, Hong Kong.

Creswell, J. W. (2018). *Research design: Qualitative, quantitative, and mixed methods approaches* (5th ed.). Thousand Oaks, CA: Sage Publications.

Daskalogianni, K., & Simpson, A. (2000). Towards a definition of attitude: The relationship between the affective and the cognitive in pre-university students. In T. Nakahara & M. Koyama (Eds.), *Proceedings of the 24th Conference of the International Group for the Psychology of Mathematics Education* (Vol. 2, pp. 217–224). Hiroshima: Hiroshima University.

DeBellis, V. A., & Goldin, G. A. (2006). Affect and meta-affect in mathematical problem solving: A representational perspective. *Educational Studies in Mathematics, 63*(2), 131–147.

Denzin, N. K., & Lincoln, Y. S. (Eds.). (2005). *The Sage handbook of qualitative research* (3rd ed.). Thousand Oaks, CA: Sage Publications.

Fisher, R. J., & Katz, J. E. (2000). Social-desirability bias and the validity of self-reported values. *Psychology and Marketing, 17*(2), 105–120.

Furinghetti, F. (1998). Beliefs, conceptions and knowledge in mathematics teaching. In E. Pohkonen & G. Törner (Eds.), *The state-of-art in mathematics-related belief research: Results of the*

MAVI activities (Research report 195) (pp. 11–36). Helsinki: Department of Teacher Education, University of Helsinki.

Goldin, G. A. (2000). Affective pathways and representation in mathematical problem solving. *Mathematical Thinking and Learning, 2*(3), 209–219.

Graves, R. P. (1889). *Life of Sir William Rowan Hamilton* (Vol. III). Dublin, Ireland: Hodges, Figgis & Co.

Hannula, M. S. (2012). Exploring new dimensions of mathematics-related affects: Embodied and social theories. *Research in Mathematics Education, 14*(2), 137–161.

Harzing, A. W. (2006). Response styles in cross-national survey research: A 26-country study. *International Journal of Cross Cultural Management, 6*(2), 243–266.

Heie, H. (2002). Developing a Christian perspective on the nature of mathematics. In A. C. Migliazzo (Ed.), *Teaching as an act of faith* (pp. 95–116). New York City, NY: Fordham University Press.

Kouba, V. L., & McDonald, J. L. (1991). What is mathematics to children? *Journal of Mathematical Behavior, 35,* 105–113.

Kvale, S. (1996). *Interviews: An introduction to qualitative research interviewing*. Thousand Oaks, CA: Sage Publications.

Lalwani, A. K., Shavitt, S., & Johnson, T. (2006). What is the relation between cultural orientation and socially desirable responding? *Journal of Personality and Social Psychology, 90*(1), 165–178.

Lam, C. C., Wong, N. Y., & Wong, K. M. P. (1999). Students' conception of mathematics learning: A Hong Kong study. *Curriculum and Teaching, 14*(2), 27–48.

Law, H. Y., Wong, N. Y., & Lee, N. Y. L. (2012). A study of espoused values in Hong Kong's mathematics classrooms. *ZDM—The International Journal on Mathematics Education, 44*(1), 45–57.

LeCompte, M. D., & Preissle, J. (1993). *Ethnography and qualitative design in educational research* (2nd ed.). San Diego: Academic Press.

Leder, G. C., Pehkonen, E., & Törner, G. (Eds.). (2002). *Beliefs: A hidden variable in mathematics education?*. Dordrecht: Kluwer Academic Publishers.

Leu, Y.-C., Chan, Y. C., & Wong, N. Y. (2015). The relationship between religious beliefs and teaching among mathematics teachers in the Chinese mainland, Taiwan and Hong Kong. In L. Fan, N. Y. Wong, J. Cai, & S. Li (Eds.), *How Chinese teach mathematics: Perspectives from insiders* (pp. 653–699). Singapore: World Scientific.

Lim, C. S. (2010). Using NVivo 8 to explore photo-voice. In J. Lim (Ed.), *Proceedings of the international workshop on computer aided qualitative research Asia* (pp. 48–59). The Netherlands: Merlien Institute.

Marton, F., & Booth, S. (1997). *Learning and awareness.* New Jersey, NJ: Lawrence Erlbaum Associates Inc., Publishers.

McLeod, D. B. (1992). Research on affect in mathematics education: A reconceptualization. In D. A. Grouws (Ed.), *Handbook of research on mathematics teaching and learning (A project of the National Council of Teachers of Mathematics).* New York: Macmillan.

Philipp, R. A. (2007). Mathematics teachers' beliefs and affects. In F. K. Lester Jr. (Ed.), *Second handbook on research on mathematics teaching and learning (A project of the National Council of Teachers of Mathematics)* (pp. 257–315). Charlotte, NC: Information Age Publishing.

Punch, K. F. (1998). *Introduction to social research: Quantitative and qualitative approaches (Essential resource books for social research).* Los Angeles, CA: Sage Publications.

Rubin, H. J., & Rubin, I. S. (2012). *Qualitative interviewing: The art of hearing data* (3rd ed.). Thousand Oaks, CA: Sage Publications.

Seah, W. T., & Andersson, A. (2015). Valuing diversity in mathematics pedagogy through the volitional nature and alignment of values. In A. Bishop, H. Tan, & T. Barkatsas (Eds.), *Diversity in mathematics education: Towards inclusive practices* (pp. 167–183). Switzerland: Springer.

Seah, W. T., & Wong, N. Y. (2012). Thematic issue on 'Values in East Asian Mathematics Education—The Third Wave'. *ZDM—The International Journal on Mathematics Education, 44*(1), 1–2.

Silverman, D. (2001). *Interpreting qualitative data: Methods for analyzing talk, text and interaction* (2nd ed.). London: Sage Publications.

Wiegerová, A. (2013). Teacher journal as a research instrument (research—the first year in the life in primary school through the eyes of novice teachers). *Studia Edukacyjne, 27,* 237–252.

Wong, N. Y. (1993). The psychosocial environment in the Hong Kong mathematics classroom. *The Journal of Mathematical Behavior, 12,* 303–309.

Wong, N. Y., & Watkins, D. (2001). Mathematics understanding: Students' perception. *The Asia-Pacific Education Researcher, 10*(1), 41–59.

Wong, N. Y., Marton, F., Wong, K. M., & Lam, C. C. (2002). The lived space of mathematics learning. *Journal of Mathematical Behavior, 21,* 25–47.

Wong, N. Y., Lam, C. C., Wong, K. M., Han, J. W., & Wong, Q. T. (2003). A study on the conception of mathematics among Hong Kong mathematics teachers [in Chinese]. *Journal of Mathematics Education, 12*(2), 2–9.

Wong, Q. T. (2003). *The conception of mathematics among Hong Kong mathematics teachers* [in Chinese]. Unpublished MPhil thesis, The Chinese University of Hong Kong, Hong Kong.

Wong, Q. T., Wong, N. Y., Lam, C. C., & Zhang, Q. P. (2009). Beliefs about mathematics and effective teaching among elementary mathematics teachers in Hong Kong. In J. Cai, G. Kaiser, B. Perry, & N. Y. Wong (Eds.), *Effective mathematics teaching from teachers' perspectives: National and cross-national studies* (pp. 217–234). Rotterdam: Sense Publishers.

Zhang, Q. P. (2010). *Mathematics teachers' professional knowledge, beliefs and their implications on their teaching* [in Chinese]. Unpublished PhD thesis, The Chinese University of Hong Kong, Hong Kong.

6

Scanning and Scoping of Values and Valuing in Mathematics Education

Philip Clarkson, Wee Tiong Seah and JeongSuk Pang

Abstract Although the ideas of values and valuing have been totemic notions in education for forever, when applied to mathematics they become quite problematic. Even today for many mathematic teachers and learners, mathematics is a value-free space. For them, school mathematics is learning the skills of manipulating numbers before moving to the more abstract ideas of algebra, and occasionally delving into geometry and measurement ideas. Likewise teaching mathematics in schools is ensuring students get good marks on the tests and examinations using whatever pedagogical techniques ensure this. Although in schools this is still the prevailing attitude to mathematics, nevertheless for some decades there has been a growing counter position in mathematics education research that problematizes and challenges this orthodoxy. It has argued that at a fundamental level there are mathematical values that underpin the doing of mathematics, and indeed the same is true for mathematics pedagogy. This chapter briefly explores a number of these notions as it introduces the various chapters in this volume.

Keywords Values · Valuing · Mathematics education · Mathematics · Mathematics teaching · Mathematics learning

P. Clarkson (✉)
Australian Catholic University, Locked Bag 4115, Fitzroy MDC, VIC 3065, Australia
e-mail: Philip.Clarkson@acu.edu.au

W. T. Seah
The University of Melbourne, Melbourne, VIC, Australia
e-mail: wt.seah@unimelb.edu.au

J. Pang
Korea National University of Education, Cheongju, South Korea
e-mail: jeongsuk@knue.ac.kr

Introduction

Values and valuing, particularly in the context of doing, teaching and learning school mathematics, is a confusing, complex, and contested terrain where colleagues have struggled to even find common definitions of values and valuing, although many have nuanced each other's ideas. This has led to an area of research that many shy away from since it seems that no one really knows what is of worth in the discussion. And yet clearly the notions of values and valuing have been at the heart of education ever since it was, and hence must have a role in researching, teaching and learning mathematics.

So, even though there is ferment in this discussion, over the years the debate regarding values and valuing and mathematics education has provided fertile intellectual opportunities for some scholars to tweak, critique, alter, expand, deconstruct, or devise their own particular standpoints. One way to explore this crucial area of scholarship is to note the contributions from various colleagues who have influenced our thinking.

Among the most influential in the context of mathematics education for many of us has been the contributions of Alan Bishop, to whom this book is dedicated. Bishop has always recognized the importance of the interaction of the social and political within mathematics education, but during his three months sabbatical visit with Glen Lean at the Papua New Guinea University of Technology in 1977 these notions crystalized into an imperative in his research. Mainly from that experience he developed his seminal book on mathematical enculturation (Bishop 1988) where among other ideas he formulated his notion of three pairs of complementary mathematical values and how they are, whether we recognize this or not, a critical influence on how we think about teaching mathematics and how students learn mathematics (Clarkson and Presmeg 2008). Bishop has written extensively on this theme ever since, and many authors in this volume use these six values as either a starting point for their research, or nuancing them, or in other ways reference them.

In the last decade or so, two main thrusts of researching values and valuing in mathematics pedagogy could be identified. The first is made up of different research studies conducted by Turkish researchers based at a number of Turkish universities. These include Büşra Kirez, Esra Selcen Yakıcı-Topbaş, Fatma Nur Aktaş, Gülcin Tan-Şişman, Nesrin Özsoy, Yüksel Dede, and many others. Amongst them, Yüksel Dede's attachment in Germany had also facilitated comparative studies between German and Turkish students and teachers. Some of these comparative studies were also conducted as part of Turkish participation in the second thrust of values research in mathematics education.

Unlike the Turkish research focus which represents the first thrust, the second thrust was not country-specific. Called the 'Third Wave Project', it is a series of values research studies conducted by as many as 23 research teams based in 20 different economies, and coordinated by Wee Tiong Seah who is based in Australia. Each study in the Third Wave Project would typically be conducted in different education systems across the world, thus producing findings at the levels of both individual economies and across economies. Amongst these, the largest scale and arguably the most well-known study would be the 'What I Find Important (in my mathematics learning)' study, commonly called the WIFI Study. More than 18,000 student questionnaires have been collected and analysed. WIFI data can be seen in Chapters. (Ghana) and (mainland China) in this book. It is important to note that for most of the participating economies such as Australia, Ghana and Japan, the coordinating institutions subsequently lead ongoing research efforts into values in mathematics education within their respective economies.

Although Bishop (1988) had conceptualised three pairs of complementary mathematical values, that is, values which characterise the nature of mathematics as it is presented in classrooms in the 'West', empirical studies that had been conducted since have not identified any other mathematical value to add to this list. At the same time, research studies have also identified different mathematics educational and general educational values, using Bishop's (1996) categorisations, as we will see in the rest of this book. These seem to span across a large range of possibilities, however.

Formation of the Book

This volume is a contribution to the book series that had its foundation at the ICME-13 conference (Kaiser 2017), held in 2016. During that conference two of the editors, Clarkson and Seah (with Alan Bishop, Penelope Kalogeropoulos and Annica Andersson), led a Discussion Group entitled *Connections Between Valuing and Values: Exploring Experiences and Rethinking Data Generating Methods* (Clarkson et al. 2017). During the first of two sessions that the Discussion Group held, we had a number of contributions that discussed where the study of values and valuing in mathematics education had come from and what was happening around the world at that time in various studies. Two key references for this discussion were a special issue of *ZDM Mathematics Education* (Seah and Wong 2012), and an article that some of us had written (Seah et al. 2016). In the second session of the Discussion Group we directed a role play, based on an earlier version which three of us had run previously at the 35th PME conference held in 2015 (Clarkson 2015), which further explored the mathematical values that Bishop had written about. This present volume extends further the discussions we held during session one of the ICME Discussion Group, and through the agency of the role-play were extended in the second session. Our call for chapters for this volume was broadcast as follows:

Despite the money and time that have been invested over the last few decades in mathematics educational research, improvements of the learning and teaching of mathematics in schools does not appear to have kept pace. For example, although there is some variation between countries, overall student misconceptions of various mathematical concepts remain the same, student engagement in mathematics learning has remained low, and we do not seem to be able to improve student attitudes towards mathematics learning.

Reasoning and feelings are clearly part of the learning and teaching of mathematics. But also involved are students' interaction with their cultural setting and each other. The relative recent development of the socio-cultural approaches to understanding and facilitating mathematics education research has complemented the more traditional cognitive and affective traditions. But added to all of these the construct of values has been a promising and useful notion. The significance of values and valuing in other fields of studies, such as science, medicine and business is well established. However values and valuing in the context of mathematics learning and teaching has only been explored from the mid 1980s.

This volume will bring together some of the world's leaders in this aspect of mathematics educational research, who will be reporting on the latest academic knowledge of a chosen theme (e.g. engagement, special education) from the values perspective, discussing how a values/valuing perspective can better facilitate a more effective mathematics pedagogical experience, and proposing implications for research and classroom practice relevant to the them. Reflecting the socio-cultural nature of the values construct, this volume will also feature the intellectual work of researchers from different ethnicities and nationalities. As a collective whole, this should stimulate the reader to further consider each of the featured themes in cross-cultural ways.

The intention of this volume then will focus on conceptual aspects, in terms of how values and valuing play a role in complementing cognition and affect in mathematics learning and teaching. In due course we hope to edit a second volume that will extend this discussion by focusing more on the practical, intervention aspects of values in mathematics education.

Chapter Outlines

We have not tried to squeeze the chapters for this volume into set sections. Although there are some obvious overlap between some chapters over and above them dealing in some way with values and valuing and mathematics, there are also some isolates. In any case the overlaps that we as editors may see might not capture important synergies, and indeed might obscure other overlaps that could be crucial to the thinking of some readers. Hence we have thought about this set of chapters more like small streams that have gradually coalesced into a larger flow, but we leave it to you the reader to decide when the various convergences happen.

With these thoughts in mind, and noting that the order of such is somewhat random, we turn now to introduce the various chapters in this volume.

Clarkson's chapter documents the recorded conversation between himself and Alan Bishop—arguably the father of research into values and valuing in mathematics education—just prior to Alan's return to UK after 25 years in Australia. As would be expected, much was discussed, which included reflections about two projects they co-led; the 'Values and Mathematics Project (VAMP)' and as the 'Mathematical Well-being construct' project. There is also a beautiful analogy of mathematics as weaving, with the warps representing the values that are inherent in the discipline. The main focus of the chapter, however, is on the three pairs of complementary mathematical values, which Bishop (1988) conceptualised. The much-talked about question over the years, 'should there be only six mathematical values?' was also touched on. Many research questions were raised throughout the conversation, either explicitly or otherwise, and these should provide many researchers with stimuli for formulating research in this aspect of mathematics education. Of course, the answers to some of these questions might be found in the other chapters of this book (e.g. 'changes in students' values' is examined in Dede's chapter), and the keen reader will no doubt delight in making the connections as s/he peruses the entire book. Delightfully, the chapter ends with a provocative sentence; 'one hopes the reader will also be challenged to think broadly on the notion of mathematical values, a crucial element of the foundational frame that holds what we understand as western mathematics'. This has the look and feel of a cliff-hanger ending to a movie, with a promise of a sequel to come! So, we are reminded of the existence of other kinds of mathematics, and thus, possibly other categories or types of values as well. It does look like there remains much more to what we currently understand and know of values and valuing in the context of mathematics education.

Carr presents an updated systematic literature review of values and valuing in mathematics education with a data set of 34 empirical studies published in peer-reviewed journals from 2003 onward. She provides us with a brief but concise summary of all the studies. In order to identify what has been achieved in this field, Carr explores where the research has been conducted, which stakeholders (e.g., teacher, student) are represented, what has been known regarding the development of values, and how consistent the findings of the research are. On the basis of the research findings, Carr suggests the evolving definition of values is a reflection of motivation and effort. She recommends that future research be carried out in respect to the role of conation in shaping values, the relationship between valuing of achievement and academic performance outcome, and changes in values in mathematics teaching and learning. She also calls for studies that go beyond one-off self-report questionnaires.

Corey and Ninomiya focus on teachers and the values they displayed when planning to teach mathematics, and when doing the actual teaching of mathematics. The authors discuss particular community values that Japanese teachers bring to their craft: writing detailed lesson plans, kyozaikenkyu (a planning practice), and emphasizing student mathematical reasoning in instruction. From their analysis they found eight specific values that seem to be essential to the Japanese mathematics teaching community. The study reported by the authors is part of a much larger longitudinal cross cultural study in Japan and the USA and they note some quite interesting contrasts and surprising similarities between these Japanese teachers and their USA counterparts. A number of the values identified here that are important to Japanese teachers clearly informed the notions of lesson study. It is probable that the implementation of lesson study might not be as efficacious elsewhere unless the non Japanese implementing teachers hold the same or similar values.

Andersson and Österling are concerned with the dilemma of conflicting values between the democratic actions intended in the Swedish curriculum and the most valued mathematical activities nominated by Swedish students. An active participation by students in mathematics classrooms has been politically emphasized, which the authors associate with the values of *openness*, *rationalism*, and *progress*. In contrast, the eleven- and fifteen-year-old students who participated in the WIFI (What I Find Important in learning mathematics) survey valued most, teachers' *explanations*, knowing the *times tables*, and *correctness*. The authors argue that students' valuing of such activities is to be understood as culturally and historically valued actions. They call for caution against the contradiction that democratic inclusion of students' concerns may conserve the values of *objectism* and *control* rather than *openness* and *rationalism* in mathematics classrooms.

Davis, Carr and Ampadu's work, researching what students value in Ghana, has presented the values research community with relatively rare insights into not just an African perspective, but also that of a country whose students do not perform well in international comparative assessments. The attributes which Ghanaian students value (or not) can help us understand the significance of what are valued (or not) in other countries. The main focus of Davis, Carr and Ampadu's chapter however, is on the potential for students' valuing to change as they transition from primary to junior secondary and then to senior secondary schools. To this end, the questionnaire responses from 1256 Ghanaian primary, junior high school and senior high school students suggest a gradual shift in intensity of student valuing across the school levels, reflecting a greater valuing of all but one of the seven highly valued attributes. Perhaps this shift is in part due to a greater emphasis on high stakes assessment at higher levels of schooling. An exception is the valuing of *relevance*, which appeared to be most highly valued at the primary school level. The change here can also be explained at least in part with the advent of high stakes assessment at higher levels of schooling. That is, students choosing to study mathematics at the higher secondary school levels might be too preoccupied by the need to perform in examinations that the valuing of mathematics being relevant could be diluted somewhat. On the whole, this study complements the findings of Zhang et al. (2016), which highlights how student values can and do change during the schooling years and provides examples of values change initiatives which are effective.

Hill, Hunter and Hunter explore what middle school Pāsifika students in New Zealand valued most for their mathematics learning in order to contribute more equitable and effective instruction. The authors found that the most important values espoused by the Pāsifika students were utility, peer collaboration/group-work, effect/practice, and family/familial support. Among these, the values of peer collaboration/group-work and family/familial support were identified as specific to the Pāsifika students. The authors argue that culturally responsive mathematics experiences have the potential to produce more equal learning outcomes for target students. As such, this chapter suggests that for equitable mathematics teaching we need to develop the classroom culture and pedagogy in a better way to align with the mathematics educational values of minority students.

Kalogeropoulos and Clarkson focus on the value alignment strategies that four Australian teachers used to enhance student engagement when critical incidents arose in the flow of mathematics teaching. Four such value alignment strategies were identified as Scaffolding, Balancing, Intervention and Refuge. These strategies supported the dynamic and flexible nature of value priorities through the ongoing social interaction in the mathematics classrooms. The authors then explore the notion of mathematical identity, which also affects engagement in mathematics learning. Given the complex interplay between values and identity, this chapter suggests that mathematical identity be considered in the alignment of values between a teacher and students.

Abdullah and Leung's chapter demonstrates the relevance of educational values in the consideration of lesson study cycles across cultures. They highlight an earlier observation by Fujii (2014) that the failure for successful replications of the Japanese lesson study model outside Japan could be attributed to a corresponding failure for relevant educational values embedded in the Japanese model to be recognised and represented in the overseas adaptations. They specifically report on a lesson study in a Brunei primary school and found that the values espoused by the teachers reflected local cultural factors (such as the bilingual learning context), which in turn affected the form in which lesson study format took in Brunei.

Dede's chapter reminds us that not only are comparative studies useful with which to study values, but the examination of values across cultures and cultural groups will also help us conduct better comparative studies. Here, Dede drew on the analysis of interviews he conducted with 35 Grade 9 students comprising of German students in Germany, Turkish students in Turkey, and Turkish immigrant students in Germany. Values unique to each group of students were identified. At the same time, it was also observed that the Turkish immigrant students in Germany were not valuing *fun* in mathematics learning, which highlights how a student's personal experiences might possibly affect how s/he views/regards school education. Dede also identified two values that were common to all three groups of students, namely, rationalism and relevance. That these two values had also been identified by students in other studies elsewhere (as cited by Dede) highlights their pan-cultural significance in mathematics learning.

Nakawa discusses the possibility of incorporating the framework of mathematical, social, and personal values into a number activity. Her qualitative analysis showed that Japanese kindergarten children regarded equality and fairness as very important among their personal and social values. Interestingly these social and personal values became a driving force leading to mathematical values. She suggests that given appropriate situations, social and personal values can serve as a catalyst for quite young children to organically develop mathematical values.

In another study set in Japan, Baba and Shimada are concerned with the notion that mathematics is perceived by students as being value free. They use socially open-ended problems to explore both social and mathematical values with primary aged students and show with such problems, teachers can explore with their students the linkages between the two. Interestingly when students considered the social implications and values inherent in a game-playing situation, they generated various mathematical models and then argued that their consideration of the inherent social values concerning the players of the game became a reason for choosing a particular mathematical model. This in itself is an interesting outcome in that so often the mathematical is privileged and other considerations, including the social values of the context, become subservient. Nevertheless the authors argue that the mathematical models that students formulated had to conform to the rationality of mathematics and they found that having students enter into dialog helps this. Indeed students were willing to modify their decisions on both mathematical and social values if they were convinced by the arguments proffered by their peers.

Zhang's chapter takes the reader to mainland China and reports on a comparison of primary and secondary students' valuing. As have other studies reported in this book, he also found that as students progress through years of schooling so their valuing also changes. He also noted that there were some differences attributed to gender. But most interesting was the insight that these students seem to prefer a teacher led approach to teaching, but at the same time a student-centred learning classroom. Such a result would seem strange if reported from western classrooms where these two often are deemed to be in opposition. Results such as these demand deeper thought be given regarding the cultural influences on both teachers' and students' valuing.

Chan and Wong provide an overview on their own previous investigations of students' and teachers' beliefs and values about mathematics, mathematics learning and teaching. The main focus of such an overview was to highlight various research methodologies employed in the studies such as open-ended questions, episode writing, hypothetical situations, mind maps, a variety of interviews (e.g., clinical, semi-structured, or stimulus-recalled), quotes from famous mathematicians, classroom observation, snapshots of critical moments, teachers' journals, and questionnaires. The authors argue for the use of hypothetical situations in which the participants need to make a choice under a dilemmatic or extreme situation, because such situations force them to reveal their values. The use of hypothetical situations is promoted to be a complementary methodology in value research to other frequently used ones such as questionnaire and interviews.

The book ends with another study from Turkey. Aktaş, Yakıcı-Topbaş and Dede report on the values that in part played a role in teachers' decision making as they were teaching about polygons in primary schools. As did Bishop and Whitfield (1972) many years ago, the authors note that teachers make decisions before, during and after a lesson. All are important but in different ways, and each set of decisions can be influenced by values (see also Borko et al. 2008). However this study focuses more on the moments during the act of teaching when critical decisions are made and which students notice. These are analyzed and the authors draw out a list of the values they noted in the lessons, that they then set into a frame to show possible inter relationships.

Beginning

This chapter is just a beginning and it facilitates an opening to the following chapters. Hence we use the continuous form for this section heading. This notion of the continuous is important in both writing and reading. In writing we may well start with a traditional format of knowing what we want to say and have a good idea of the beginning and where and how we want to finish. But in between strange things can happen. To our amazement in writing, a new emerges and the world as we know it changes. In some way we write and discover new ideas and notions about our selves, and so we grow. And it can be the same in reading: We can start by reading a chapter, perhaps because we think it will say something important about ideas we have already formulated, but it does not always work that way. At times, and we hope this happens through the agency of this volume, we end up in quite unexpected places.

This book features the intellectual work of researchers from different ethnicities and nationalities. We hope it will be thought provoking and will be stimulating for you the reader just as much as it has been for us the editors when we put this volume together. As well we hope that the volume will provide an impetus for future conversations about mathematics and values and valuing as we struggle with the immense and ongoing controversies and challenges to make sense of this area, which will bring further insights to researchers, teachers and the wider community.

References

Bishop, A. (1988). *Mathematical enculturation: A cultural perspective on mathematics education*. Dordrecht: Springer.

Bishop, A. J., & Whitfield, R. (1972). *Situations in teaching*. London: McGraw Hill.

Bishop, A. J. (1996). How should mathematics teaching in modern societies relate to cultural values—some preliminary questions. In *Paper presented at the seventh southeast asian conference on mathematics education*. Hanoi, Vietnam.

Borko, H., Roberts, S., & Shavelson, R. (2008). Teachers' decision making: From Alan Bishop to today. In P. Clarkson & N. Presmeg (Eds.), *Critical issues in mathematics education: Major contributions of Alan Bishop* (pp. 37–67). Dordrecht: Springer.

Clarkson, P. (2015). Discussion Group report: Connections between valuing and values; rethinking data generating methods. *PME Newsletter* 7–10.

Clarkson, P., Andersson, A., Bishop, A., Kalogeropoulos, P., & Seah, W. T. (2017). Connections between valuing and values: Exploring experience and rethinking data generating methods. In G. Kaiser (Ed.), *Proceedings of the 13th International Congress on Mathematical Education ICME-13* (pp. 643–644). New York: Springer.

Clarkson, P., & Presmeg, N. (Eds.). (2008). *Critical issues in mathematics education: Major contributions of Alan Bishop*. Dordrecht: Springer.

Fujii, T. (2014). Implementing Japanese lesson study in foreign countries: Misconceptions revealed. *Mathematics Teacher Education and Development, 16*(1), 65–83.

Kaiser, G. (Ed.), (2017). *Proceedings of the 13th International Congress on Mathematical Education ICME-13*. New York: Springer.

Seah, W. T., Andersson, A., Bishop, A., & Clarkson, P. (2016). What would the mathematics curriculum look like if values were the focus? *For the Learning of Mathematics, 36*(1), 14–20.

Seah, W. T., & Wong, N.Y. (Eds.). (2012). Values in East Asian mathematics education – The third wave (Special edition). *ZDM – The International Journal on Mathematics Education, 44.*

Zhang, Q., Barkatsas, A., Law, H., Leu, Y., Seah, W., & Wong, N. (2016). What primary students in the Chinese mainland, Hong Kong and Taiwan value in mathematics learning: A comparative analysis. *International Journal of Science and Mathematics Education, 14*(5), 907–924.

7

Mathematical Values through Personal and Social Values: A Number Activity in a Japanese Kindergarten

Nagisa Nakawa

Abstract This chapter discusses the possibility of incorporating the framework of mathematical, social, and personal values into a number activity, following the approach shown in Shimada and Baba (Transformation of students' values in the process of solving socially open-ended problems. In: Beswick K, Muir T, Wells J (eds) Proceedings of 39th psychology of mathematics education conference, vol 4. PME, Hobart, Australia, pp. 161–168, 2015). The qualitative analysis showed that kindergarten children regarded equality and fairness as very important among their personal and social values when distributing sweet potatoes to two different kinds of animals in an imaginary restaurant. Further, these social and personal values became a driving force toward mathematical values: children verbally expressed their logical opinions related to different quantities (the size of animals' mouths), an ability closely linked with the beginning of rationality in mathematical values. Therefore, social and personal values can serve as a catalyst for kindergarteners to organically develop mathematical values.

Keywords Social values · Personal values · Mathematical values · Mathematical play · Kindergarten mathematics

Introduction

In Japanese public kindergartens, which are pre-schools for children aged three to six, cultivating a foundation for lifelong character-building through play is viewed as crucial (Japanese Ministry of Education, Culture, Sports, Science and Technology [MEXT] 2008, p. 1). Teaching and learning in Japanese public kindergartens are not subject-based, but play-based, connecting to children's daily life to foster their mental and physical development. Therefore, mathematics is not a discrete "subject"; instead, mathematical skills are integrated across the five focal areas to be learned

N. Nakawa (✉)
Kanto Gakuin University, 1-50-1, East Mutsuura, Kanazawa Ward, Yokohama, Kanagawa 236-8501, Japan
e-mail: nagisa@kanto-gakuin.ac.jp

in the national Course of Study for Kindergarten: health, human relationships, environment, languages, and expression. These are said to be essential for children's cognitive and physical development at pre-primary level (MEXT 2008, 2017b). The ultimate objective of kindergarten in Japan is to establish a foundation for the formation of one's mature personality (MEXT 2017b, p. 3). The author and the research team involved in this developmental project for early mathematics, surmised that teaching mathematics including different values would fit into kindergarten education if it included an element of values education.

This chapter reports on the implementation and results of that kindergarten mathematics project, with a particular focus on different values in a mathematical activity—the Development of the Elementary Mathematics Education Training Program for Preschool Teachers and Parents (the DEMETP). Focusing on mathematics, which deals with children's values as well as their cognitive development in preparation for primary mathematics, the goal of this chapter is to use the perspective of children's social, personal, and mathematical values to examine the quality of learning about numbers through a teaching/learning activity called "sweet potato sharing".

DEMETP Project

The aim of the DEMETP project is to develop a curriculum for mathematics teaching and learning which fits into the current Course of Study for Kindergarten. As Matsuo (2015) has explained, the project broadly focuses on the formation of relationships between teacher and children and on interactions between the children and objects in mathematical activities. The project has three pillars in its curriculum framework. The first is the classification of educational objectives in relation to the cognitive and affective development of children. The second is mathematical content and teaching method. While primary school mathematics in Japan consists of four areas—numbers and calculation, quantity and measurement, geometry, and the relationship between number and quantity—this project broadly covers number and quantity, shapes, number change, and the relationship between numbers and quantity to link kindergarten and primary school mathematics. It is concerned with basic problem-solving and representation considering the developmental stage of young children. The third pillar integrates the five areas to be learned as mentioned in the Course of Study for Kindergarten with basic learning attainment and mathematical literacy. The stages of each activity in this project are as follows: (1) adopting an activity and setting a problem, (2) preparation, (3) organisation of the classroom, and (4) implementation of the activity. So far, the project team has developed nineteen activities in different areas of mathematics and are in the process of examining their effectiveness.

Learning Numbers and Division

Mathematics is barely discussed in the Course of Study for Kindergarten. As mentioned, mathematics is not a separately-taught subject; however, children are encouraged to learn mathematics-related concepts to understand numbers, quantities, and shapes in ways that can be integrated into the five focal areas. On the other hand, in the new Course of Study for Kindergarten, an interest in and a sense of numbers, quantities, and shapes are emphasised as one of the expectations of how children should grow at the end of kindergarten education, which was not mentioned explicitly in the previous version. Further, under the Environment Objective, the importance of mathematics is mentioned thus: "Pupils should have an interest in numbers, quantities, and shapes in their daily life" (MEXT 2017b, p. 15). Other than this statement, there is no clear reference to mathematics.

It is in Grade 1 of Japanese primary school, when the children are aged six, that the basic concepts of numbers and numerals begin to be officially and formally dealt with. In primary school mathematics, children are first expected to learn how to count, read, and write numerals up to ten, understand the structure of numbers, and use numerals in reference to concrete and semi-concrete objects. Next, they start learning to compose/decompose numbers up to ten, in preparation for the introduction of addition and subtraction (MEXT 2017a). Children learn multiplication and division for the first time in Grade 2 (age seven–eight). However, some studies (Fuson 1992; Maruyama 2004) suggest that younger children can also manipulate numbers in certain ways, such as "counting on" (Fuson 1992, p. 121), before entering primary school. In Japanese kindergartens, children also informally experience the use of numbers in their daily life; for instance, counting concrete objects, reading numerals, paying money, figuring out which object relates to another given object, and so on. The team included these mathematical concepts in the kindergarten activities in this project.

Compared with primary school mathematics, kindergarten has three peculiar features. First, children learn through playing (Cohrssen et al. 2014; Vogel 2013; Thomas et al. 2011). Second, it is more significant for them to experience mathematical play, and it does not really matter if their answers are correct. Trial-and-error takes primacy at this stage of mathematics education. Finally, kindergarten children deal with concrete material in the ikonic mode and those who are in primary school move to the semiotic mode, where they start using mathematical symbols, per the Structure of Observed Learning Outcome (SOLO) model (Pegg and Tall 2005). This is a structural model of cognitive development that defines a sequence of cognitive levels and cognitive development modes according to the children's ages. This model can help us identify the degree of children's understanding.

In this project, therefore, the team developed activities on the structure of numbers, cardinal and ordinal numbers, and the basic ideas of dividing concrete objects—including the composition and decomposition of numbers. These activities were intended to foster a smooth transition to primary school mathematics. This chapter focuses on dividing concrete objects, related to composition/decomposition.

Values

The Course of Study for Kindergarten states the importance of values such as collaboration, empathy, and adhering to a moral standard and social norms (MEXT 2017b, pp. 4–5). Its environment and human relationship objectives include the following on *values*: children should share the various feelings of joy and sadness with the teacher and other children in groups/class and through active interactions with others, think independently and decide to act, convey what they feel and listen to others, and have an interest in and concern for the environment and nature around them (MEXT 2017b, pp. 13–15). These descriptions explain this project's interest in values in mathematical activities for kindergarten.

Research on values in mathematics education (e.g. Kalogeropoulos and Bishop 2017; Seah et al. 2017; Zhang and Seah 2015; Shimada and Baba 2015; Seah et al. 2001) generally proposes three types of values as relevant to mathematical learning: *mathematical values*, *mathematics educational values*, and *general educational values* (Bishop 1996). However, Japanese mathematics educators Shimada and Baba (2015), introduced a different tripartite division: *mathematical values*, *social values*, and *personal values*. They focused on students' social values and their transformation through problem-solving in class at the primary level in Japan and discussed the fundamental question of whether children at the kindergarten level exhibit social values. According to them, mathematical values include rationalism (Bishop 1988), while social values are defined as notions that children present in real-life settings and society in problem-solving (Shimada and Baba 2015). Bishop (1988) and Shimada and Baba (2015) both defined mathematical values in a similar manner. Considering the ultimate objective of kindergarten in Japan, social and personal values and their development are regarded as important for children's general development, with mathematical and mathematics educational values less so. This chapter will adopt Shimada and Baba (2015)'s framework. It will also examine how both social and personal values relate to mathematical values, which are fundamental values that children will carry into mathematics class in primary school. Successfully capturing these three types of values during mathematical activities at the kindergarten stage will provide even more reason to focus on mathematical activities as they foster a variety of values in children while offering fundamental values that children require in order to be good citizens.

Considering the developmental stage of kindergarten-age children, it is likely to be difficult for them to communicate their values through discussion. However, it may be possible for the teacher to create a setting in which they can show or manifest their own social and personal values, which are also connected to their mathematical values. It is worthwhile to examine children's values in mathematical learning contexts because values-related mathematical activity offers an opportunity for children to share their ideas and thinking. In other words, mathematical activities planned with values in mind may integrate two different skills: deepening the basic concepts of numbers and division, and expressing one's opinion based on one's daily life and ideas. Both of these seem essential to learning at the kindergarten level.

Method

The author developed a mathematical activity called "sweet potato digging" in the framework of this program.

(1) *Setting a problem within the activity:* Sweet potatoes have long been popular in Japanese agriculture and related to cultural life in Japan since the Edo period (B.C. 1600) (Ito 2010). Children plant and harvest them with teachers' support for educational purposes as a seasonal and special activity. They plant sweet potatoes around May (which can also be an opportunity for other types of learning, such as learning about agriculture). The activity can continue right through to harvesting in October. On the day before the final part of this activity, the children go out into the field and dig up the sweet potatoes. On the following day (the day of the activity), they wash the sweet potatoes in the morning and then engage in an activity about composition/decomposition of numbers using them. After the activity, they eat the sweet potatoes. Thus, the activity is connected to the children's daily life—as both a familiar food and learning tool—and to seasonal events such as planting and harvesting.
(2) *Preparation/materials needed:* Sweet potatoes, pictures of mice and moles, and rubber bands.
(3) *Purpose of the activity:*

    (i) To find pairs of numbers using sweet potatoes that add up to five or to ten, and to recognise that these numbers always add up to the same number—five or ten, and;
    (ii) To observe what kind of values are appearing throughout the activity.

(4) *Class/group organisation:* Team teaching (one teacher teaches while the other supports the activity when needed) and pair activities. The teacher chose an activity in which students were paired due to time constraints, as well as because the resultant collaboration and verbal communication between pupils are significant for learning.
(5) *Implementation of the activity:*

    (i) To reflect on sweet potato digging on the previous day and to understand the setting of the activity.
    (ii) To introduce mice and moles and explain that the children are going to feed them.
    (iii) To decompose five: Children in each pair put rubber bands on sweet potatoes and decide how many will be given to mice and moles respectively, placing the rubber bands on the sweet potatoes and putting them into a designated box.
    (iv) To explain how and why they decided to distribute the sweet potatoes in such a way.
    (v) To make sure the sum of the two numbers is five.

(vi) To decompose ten: Representative pairs, who are randomly chosen by the teacher, place rubber bands on sweet potatoes and decide how many sweet potatoes are going to mice and moles respectively and put them into the designated box.
(vii) To explain how and why they decided to distribute the sweet potatoes in such a way.
(viii) To make sure the sum of the two numbers is ten.
(ix) To reflect upon the activity.

Results

The activity was conducted in a public kindergarten in Tokyo in October 2015. After the research was permitted by the university, the author was given permission to take photos of children by the school principal and class teacher. The author recorded the activity for data collection and took fieldnotes of the different group activities as each group was working at the same time. The activity consisted of three separate trials. Two teachers, the class teacher and head teacher who implemented this activity, prepared four desks at which the pairs worked; the rest of the children awaited their turn, sitting on the chair and listening to the presentations after each trial.

After the activity, the DVD data were transcribed and analysed qualitatively. The number of participating children was sixteen, aged five to six. In class, the teacher created a backstory to help the children understand the activity and imagine a context for it. She told them about an imaginary restaurant that served sweet potatoes to moles and mice, and asked the children to decide how many to serve to these "customers", supposing that the two groups of animals could eat five altogether. She also asked them if they could find novel ways to divide the sweet potatoes—one for the moles and four for the mice, for example; or maybe three for the moles and two for the mice.

Following this explanation by the teacher, every child understood what they were supposed to do. Then, a mathematically bright child asked:

> 1 S1: Can we divide one sweet potato in half, because we want to divide them equally to both mice and moles? I would feel very sorry for the mice, if mice were given two, and moles were given three.

To which the teacher replied:

> 2 T: Because we do not have any knives to cut, we cannot halve one sweet potato, can we? But I understand that you want to split them equally, right? If moles were given more, the next pair might give more sweet potatoes to mice than the first group.

This child's question shows that he thought it was important to divide things equally. After the conversation, eight pairs of students completed the activity, dividing the sweet potatoes among the mice and the moles in various ways across the three trials. Table shows the results of the activity, including conversations and actions.

Table: How each pair divided the sweet potatoes and why

Pair no.	No. of sweet potatoes		Reasons, talk, or actions for dividing in this way, according to children
	Mice	Moles	
1	3	2	"Because mice are smaller than moles, we feel sorry for the mice"
2	2	3	(No response)
3	1	4	"Because this (a sweet potato for the mice) is too big compared to these ones for moles"
4	2	3	"Because we feel sorry for the small mice"
5	3	2	(Child A) "Because the mice are smaller" (Child B) "Because the mouths of moles are very big, and the mouths of mice are small"
6	3	2	(Teacher) "You gave the bigger size of sweet potatoes to moles, and the smaller size to mice?" (Children) "Yes." (The children were not able to express the reason although the teacher asked them.)
7	5	0	(The teacher told the other children instead of this pair.) "They thought they wanted to try a way of dividing which had never happened so far, right?" (The pair nodded. They could not express their reason but agreed with the teacher by nodding.)
8	4	1	"We gave more to the mice because they were hungry"

Table shows that the most common way that these children choose of dividing five sweet potatoes was into two and three. There were a few groups such as pairs 6 and 7, which were not able to say anything, and the teacher helped them out, so they ended up agreeing with her. It also reflects the children's own logical thinking as expressed by them verbally: the reasons for the division and distribution were related to the size of the animals and their mouths. For instance, pair 6 distributed three sweet potatoes to mice and two sweet potatoes to moles, but carefully chose the bigger sweet potatoes for the moles. Two pairs 3 and 8, chose to divide the sweet potatoes into one and four, with a focus on the size of the animals. The seventh pair, which decided to divide them five/zero, was strongly criticised by the other children as shown in the excerpts from the class. This conversation was as follows:

1 T: Well, next (pair 7 in Table), let's have a look. Wow, this is new. They want to create a new approach. They seem to have given only mice sweet potatoes. Can you count how many you have given to the mice, together?

2 S1: One, two, three, four, five.

3 T: I see, I mean, how many were given to the moles?

4 S2: Zero.

5 T: Wow, zero. Mice are given five and moles are given zero. How many sweet potatoes do they have altogether?

6 S1: Five (Showing five using their fingers too).

7 T: Wow, moles were not given any sweet potatoes to eat, even when we have found five sweet potatoes.

8 S3: The moles are so sorry.

9 S4: They are pitiable. (Noise from some children: "Pitiable, pitiable!")

10 S5: I feel pity.

11 T: Well, I would guess that they might think they want to create something new that nobody has served before in the restaurant. Okay? And thank you very much to the seventh group.

As the transcript shows, while the teacher was interested in and respectful of the seventh pair's mathematically novel approach, the other children were not happy at all and were generally more attentive to fairness than to new mathematical findings. Pair 7 kept quiet, but it seemed that they felt relieved by the teacher's support as everyone in class was complaining about their result.

After the pair activity, the class proceeded to the next task, which involved ten sweet potatoes. The teachers combined the two restaurants that had been running simultaneously up to that point into a bigger one, saying, "Very hungry moles and mice! We are now in a bigger restaurant." The children then counted the number of sweet potatoes, which was now ten. Two pairs of children came to the front and demonstrated how they would divide ten sweet potatoes. One pair distributed three to the mice and seven to the moles, while the other distributed four to the mice and six to the moles, as shown in the excerpts from the class interaction below:

1 T: The first and second group showed different results. They are different, aren't they? In the first pair, the moles can eat seven sweet potatoes. On the other hand, in the second pair, the moles can eat only six. This means…

2 S5: Six is less than seven by one.

3 T: Oh, yes. In the second pair, the moles are given six and the mice are given four, and then we find ten sweet potatoes. Hm? In the first pair, seven sweet potatoes are distributed to moles, three to mice. In the second pair, four to mice and six to moles, but the total number of sweet potatoes is…

4 Most children: Ten.

5 S7: Yeah, ten. (Instantly answering after 4)

6 T: I thought it would not be like that, but we found ten.

7 S8: And if we add, they will become twenty.

Discussion

Cognitive Outcome: Children's Activities of Dividing Two Quantities from a Logical Perspective

Working in pairs, all the children succeeded in dividing the sweet potatoes. While they were separating them or explaining their chosen distribution, they counted the objects by using their fingers and saying the numbers simultaneously. All of them were able to count up to five in this way. Then, two representative pairs succeeded in decomposing ten. Unfortunately, other pairs did not complete that part of the activity due to limited time, although they showed an interest in trying it. When counting numbers, everyone counted out loud. At the end of the activity, one child noted that ten plus ten is twenty, which was an advanced observation. Others were also counting from ten to twenty.

In the activity, children used concrete objects to help them with decomposing quantities, which might have been part of why they succeeded in dividing ten, and not only five. In the planning phase, the teachers were dubious about whether to introduce ten, because they felt that some children would not be able to handle the bigger quantities. Given the actual results, from a mathematical education point of view, it would be valuable to attempt this activity again to check if all students can work well with ten if given more time.

The teachers verbally confirmed the children's choices and the reasons for them, but emphasis was not placed on this aspect of the activity, and it may be that the children's thinking was not accurately reflected in some cases, such as with the group who split the sweet potatoes into five and zero and kept quiet. This case was not the precise decomposition the author expected, but the fundamental idea of addition was shown. This way of dividing did not make sense to the other children, but it is meaningful as addition. It can be conjectured that the pair 7 children operated solely in the mathematical context of numbers.

As shown in the case of ten in the above transcript, only a few children who seemed to understand the concept of bigger numbers such as ten and twenty participated or responded to the teacher's questions. Another finding was that some children understood the concept of half and double, as shown in the conversations of Sl and teacher for the first, and the last conversation of S8 in the vignettes in the previous section. Thus, various children showed a range of mathematical-cognitive ability, but all had a solid minimum threshold of understanding toward the number five, when using concrete objects.

Children's Social and Personal Values Shown Through the Activity

Although the children's values were not always clearly distinguished, with a few exceptions, the qualitative data above hints at two characteristics of these children's personal and social values. First, the children regarded equality as very important. That is, they appeared to value equality in that most pairs tried to give equal amounts to the different animals and had compensating reasons why one animal received more than the other. Historically speaking, during the Edo Era (1600–1867), the Japanese people shared the view of equality as an important philosophy even before the Meiji Restoration (around 1867). The philosophy, on the other hand, was somewhat different from what Western countries regarded as equality, according to Suzuki (2009). Earlier in Japan, equality was based on the philosophical ideas of Buddhism, while the idea of equality in Europe is embedded in Christianity. Moreover, a few philosophers, who studied in different countries in Europe, imported various ideas of equality. This actually accelerated the difference in its interpretation. In the current school education system, equality between boys and girls, as well as co-operation and sharing with friends in an equal manner, are emphasised, which Cummings (2014) also stated with regards to the Japanese primary school. Moreover, both the government and society believe that achieving equality is a fundamental and significant goal—especially between men and women, as well as between those with a disability and the able-bodied (e.g. MEXT homepage 1992; Gender Equality Bureau Cabinet homepage 2018). A similar view of what society regards as important was also observed during the activity in the kindergarten.

At the beginning of the class, a boy (S1) wanted to cut one sweet potato in half, a suggestion that seemed advanced for the kindergarten level. Here, the boy showed the personal values of equality connected to mathematical values in that he presumably believed that having equal amounts is the same as the philosophical notion of equality. However, this may not always be the case, as in societal values, other aspects of the context may override the notion of equal amounts.

Table also supports the importance of equality among children. For example, pair 3 considered the 'size' of the sweet potatoes allocated to the moles and mice, to make the actual quantities more equal, rather than the 'numbers' of sweet potatoes equal. This also reflects the social and personal values of equality. Second, the teacher wanted children to find new ways of dividing the sweet potatoes, but the children thought that fairness was very important and preferred to focus on that instead of novelty. Giving all five sweet potatoes to only one group of animals by pair 7 was not accepted by the majority of children, even though it was a new approach as the teacher suggested; the children could not understand how such division would be fair. This implies that the children were strongly engaged in the context created by the teacher, and that the social values evoked by the mice and moles were much greater than the mathematical novelty that the teacher encouraged, which will be emphasised in the children's primary school mathematics lessons. The children's valuing of the importance of equality has come from their daily life situations, where, for instance, they divided food equally with friends and siblings, which the teachers reported to us after the activity.

This also reflects the values adults convey to children in Japanese kindergarten and society and could easily lead into activity and concept acquisition related to addition, subtraction, and division in the future. The children's strong sense of fairness may be possible to harness as a driving force to help them learn how to divide equally in mathematical terms in the future. The author intends to plan and implement such an activity at the kindergarten level.

From Social and Personal Values Toward Mathematical Values

First, the boy who asked if he could halve the sweet potato, as well as pair 7 demonstrated the mathematical value of rationalism, according to the definitions of mathematical values. The first boy wanted to divide the potato equally, which is a basic fractional way of thinking, and clearly connected to mathematical values. Further, the seventh pair provided a different, novel mathematical possibility for the division of the sweet potatoes. This was partially because the teacher encouraged the children to find new ways, and they acted on this suggestion. Although the other children disagreed with the pair, their disagreement was on non-mathematical grounds. The seventh pair's action and thought are primarily based on their mathematical values. In primary school, learning often happens in daily life situations when children think mathematically. This action is connected to the beginning of these ways of mathematical thinking. Second, the relationships some students focused on were firstly the number of sweet potatoes respectively given to mice and moles, and secondly the size of the animals and their mouths. They engaged with the imaginary setting and balanced multiple variables—different relationships of different quantities and sizes—in a fairly sophisticated way, which is the very first stage of mathematical functions. Based on their verbal responses, dealing with many variables in this way can be seen to have helped elicit the start of rationalism in mathematics for them, even though their ways were primitive and sometimes needed to be assisted by the teacher. That is, the social and personal values they showed—in brief, fairness—were the driving force for them to think and express themselves logically, which was then related to the development of important elements of their mathematical values, which can be rationalism.

Conclusion

The focus on children's social and personal values in the activity helped demonstrate that they met the developmental requirements for the educational stage and to some extent, rationalism (one of the mathematical values) appeared in some scenes. At the kindergarten level, it is significant to grow their personal and social values but at the same time, this case study showed there is a possibility to foster their mathematical values. If kindergarten activities, including activities for primary education such as in the case study, contain aspects of the three given values, they will be effective both for holistic development and mathematical development in the children's lives.

Acknowledgements I would like to show my appreciation to the teachers and principal of the kindergarten where this study was conducted. The research was funded by Japan Society for the Promotion of Science research grant no. 15H02911. I also would like to thank Editage (www.editage.jp) for English-language editing.

References

Bishop, A. (1988). *Mathematical enculturation: A cultural perspective on mathematics education*. Dordrecht: Kluwer Academic Publishers.

Bishop, A. (1996). *How should mathematics teaching in modern societies relate to cultural values—Some preliminary questions*. Paper Presented at the Seventh Southeast Asian Conference on Mathematics Education, Hanoi, Vietnam.

Cohrssen, C., Church, A., & Tayler, C. (2014). Pausing for learning: Responsive engagement in mathematics activities in early childhood settings. *Australasian Journal of Early Childhood, 39*(4), 95–102.

Cummings, K. W. (2014). *Education and equality in Japan*. Princeton: Princeton University Press.

Fuson, K. C. (1992). Research on learning and teaching addition and subtraction of whole numbers. In G. Leinhardt, R. T. Putnam, & R. Hattrup (Eds.), *Analysis of arithmetic for mathematics teaching* (pp. 53–187). Hillsdale, NJ: Lawrence Erlbaum.

Gender Equality Bureau Cabinet homepage. (2018). *What is a gender-equal society?* http://www.gender.go.jp/about_danjo/society/index.html.

Ito, S. (2010). *Sweet potato and Japanese*. Tokyo: PHP New Books.

Kalogeropoulos, P., & Bishop, J. A. (2017). *What is the role of value alignment in engaging mathematics learners?* Paper Presented at the 9th Mathematics, Education and Society Conference, Volos, Greece.

Maruyama, R. (2004). Children's strategies of addition and subtraction tasks of sets. *Bulletin of Joetsu University of Education, 24*(1), 257–270.

Matsuo, N. (2015). Shaping pre-school mathematics education program: Focusing on the activity of comparing area. In *Proceedings of 48th Autumn Research Forum* (pp. 31–34). Shinshu, Japan: Japan Society of Mathematical Education.

Ministry of Education, Culture, Sports, Science, and Technology, Japan (MEXT). (1992). *Survey for equality of men and women*. Retrieved March 22, 2018, from http://www.mext.go.jp/b_menu/hakusho/nc/t19921101001/t19921101001.html.

MEXT. (2008). *The kindergarten course of study*. Tokyo: MEXT.

MEXT. (2017a). *The elementary school course of study*. Tokyo: MEXT. http://www.mext.go.jp/component/a_menu/education/micro_detail/__icsFiles/afieldfile/2017/05/12/1384661_4_2.pdf.

MEXT. (2017b). *The course of study for kindergarten*. Tokyo: MEXT. http://www.mext.go.jp/component/a_menu/education/micro_detail/__icsFiles/afieldfile/2017/05/12/1384661_3_2.pdf.

Pegg, J., & Tall, D. (2005). The fundamental cycle of concept construction underlying various theoretical frameworks. *ZDM Mathematics Education, 37*(6), 468–475.

Seah, W. T., Bishop, A. J., FitzSimons, G. E., & Clarkson, P. C. (2001). *Exploring issues of control over values teaching in the mathematics classroom*. Paper Presented at Annual Conference of the AARE, Fremantle. http://citeseerx.ist.psu.edu/viewdoc/download?doi=10.1.1.499.1752&rep=rep1&type=pdf.

Seah, W. T., Davis, K. E., & Carr, E. M. (2017). *School mathematics education through the eyes of students in Ghana: Extrinsic and intrinsic valuing*. Paper Presented at CERME10, Dublin, Ireland. Retrieved from https://www.researchgate.net/publication/312452221/download.

Shimada, I., & Baba, T. (2015). Transformation of students' values in the process of solving socially open-ended problems. In K. Beswick, T. Muir, & J. Wells (Eds.), *Proceedings of 39th Psychology of Mathematics Education Conference* (Vol. 4, pp. 161–168). Hobart, Australia: PME.

Suzuki, S. (2009). Freedom and equality in Meiji Japan. *Japan Research, 40*, 377–391.

Thomas, L., Warren, E., & de Vries, E. (2011). Play-based learning and intentional teaching in early childhood contexts. *Australasian Journal of Early Childhood, 36*(4), 69–75.

Vogel, R. (2013). Mathematical situations of play and exploration. *Educational Studies in Mathematics, 84*, 209–225.

Zhang, Q., & Seah, W. T. (2015). Chinese secondary teachers' and students' perspectives of effective mathematics teaching: The underlying values. In K. Beswick, T. Muir, & J. Wells (Eds.). *Proceedings of 39th Psychology of Mathematics Education Conference* (Vol. 4, pp. 337–344). Hobart: PME.

8

A Conversation with Alan Bishop

Philip Clarkson

Abstract We wondered, why does the mystery of mathematics seem to disappear from students: Is it because teachers have never experienced mathematical mystery? We wondered would more use of projects and investigations promote a range of mathematical values than is possible when only traditional teaching approaches are used? We wondered do teachers and students need to reach some threshold of mathematical knowledge if they are to see the inherent mathematical values that help to hold the potential disparate elements of mathematics together? These and other wonderings emerged as Alan Bishop and the author engaged in conversation that culminated some 25 years of pondering mathematical values together.

Keywords Alan bishop · Mathematics · Mathematics education · Values

Introduction

It was in 1976 when I first made contact (by snail mail) with Alan, then at Cambridge University, when I was studying for my Master of Education degree. From then on our paths occasionally crossed until Alan came to Australia in 1992 to take up a position at Monash University (Clarkson 2008a). I was by then at Australian Catholic University (Melbourne campus). Hence opportunities for working more closely together became a reality. One issue that our conversation both on and off the golf course kept returning to was values and mathematics. These notions had come into stark relief when each of us quite separately spent time in Papua New Guinea interacting with students and teachers. For Alan "it was his own experiences while living in Papua New Guinea (in 1977) that transformed (his) thinking. No longer for him were the social, cultural and political issues of some importance; they became *the* important issues with which he needed to try and come to grips, as far as teaching of mathematics was concerned"

P. Clarkson (✉)
School of Education, Australian Catholic University, Locked Bag 4115, Fitzroy MDC, VIC 3065, Australia
e-mail: Philip.Clarkson@acu.edu.au

(Clarkson 2008a). And these issues still need to be dealt with to this day (Vilson 2017).

The rest of this chapter will have excerpts of the last conversation that we had in Australia, audio-recorded in my office. Most of the references have been added later. I had compiled some 'starter ideas' with which to structure the conversation, and these are shown as figures. When contemplating these ideas before the conversation, Alan had made some notes and these appear in italic typeface. My post conversation reflections appear in plain typeface and are inserted at places as appropriate.

The 'Original' Six Values

We began our conversation by returning to the six values that Alan had used in his seminal book (Bishop 1988).

PHIL: I am talking to Alan Bishop just before he returns to England after 25 years in Australia. It is probably the last conversation we will have about values and mathematics in Australia.
ALAN: Yes I think that will be right. I have been thinking about your starter statements for a month or so. I have written out some notes that overlap with those ideas.
PHIL: I read through the starters again this morning. I guess I started thinking about the original values again after a particular conversation we had playing golf 6 months or so ago. I had forgotten about the notion of investigations and projects (Fig.), which did not feature heavily in the VAMP[1] project (Clarkson 2008b; Clarkson et al. 2010).
ALAN: I also thought about 'Starter 1' and wondered whether the six math values and their sequencing still made sense. And it still does to me. Yes 'Starter 1' made me think about quite a lot of other things. The book was much more to do with students. So we focused more on the teachers in VAMP and that still goes on.
PHIL: Well remembering back to what started our discussions that we have sustained over the years, there are still many threads to explore that come from these notions.

Starter 1: In our work of the last 25 years or so, we have not emphasized the teaching of projects, which you originally linked to the societal component of your model, as well as investigations, which you linked to the cultural component. Both teaching approaches certainly break the mould of traditional teaching in that students have to keep at the one thing for multiple lessons. That in a way has left traditional teaching approaches you associated with the symbolic component. I suppose in the VAMP project we were deliberately leaving it to the teachers to 'teach in their normal manner'. Do you still think that projects / investigations are a useful context for those four values?

Fig. Starter 1 for our discussion

[1] VAMP stands for the *Values and Mathematics Project*. This was a project Alan and I ran from 1997 through 2001. It was in part funded by small and large Australian Research Council (ARC) grants.

ALAN: Yes that's the case. My reading of the Starters has reinforced my thinking more deeply about teachers and curriculum. That's where some of the original ideas we have generated come through. The notion of the 6 values in 3 pairs; that was good:

The following are at the pedagogical level:

Projects teaching approach => societal components: control and progress
Investigations teaching approach => cultural component: openness and mystery
Traditional teaching approach => symbolic component: rationalism and objectism.
The other structural element in the book that was important are the five levels; cultural; societal; institutional; pedagogical; individual.

We then started discussing real life possibilities for teaching using investigations and projects.

ALAN: A couple of days ago I heard a very good talk about gambling. It was very interesting. I raised the question that 'does the presence of the poker machines emphasize the negative sides of gambling?' You know there is just a little step between those poker machines to other (games) machines that kids are playing with these days. How much is there a gambling factor in that? I reckon it could be a very strong factor. The challenge of the games injects a bit of competition for the kids.
PHIL: Yes
ALAN: And kids love the competition
PHIL: Playing on them (the game machines) might be sort of a bit of 'digging of the field' or preparation before they get into the gambling. And ah clearly there is always the probability that you will win but you also learn that the house will never lose. Overall the machine will never lose. The machine will always make some small profit and as time goes on that is enough for them to keep the whole circus going.
ALAN: Yes; *Bringing down the house* (Mezrich 2003). Have you read the book? Lovely book. And the movie too; *21* (Spacey et al. 2008).
PHIL: But those sorts of things get at some of the sorts of values of thinking through investigations I guess … As long as the teachers have in mind some of the mathematical values that could be taught through this, as well as societal.

Afterwards I reconsidered the issue of teaching using investigations and projects. In these teaching contexts some crucial issues need to be addressed which may not be at all obvious for young teachers, and for teachers whose confidence in teaching mathematics is low. Students do need to be given choices, not necessarily regarding the issue that will be the focus of the project or investigation, but certainly within it. But their choices will not be 'real' if students cannot make personal connections with the issue. So projects that bring out great mathematics, but really only have interest for adults, perhaps are not appropriate. Hence I am not sure that gambling would be an appropriate topic until late primary school. I never used it until early years of secondary school.

But choice of this sort needs to be balanced by reality. There is a school curriculum to teach, and possibly nation-wide testing focused on specific skills and content.

Hence although at times there might be a wide choice given to students, at other times choices may be constrained by content that must be covered in a particular time frame.

Another issue is that projects and investigations can often focus just on the answer. So primary schools students who produce wall charts or power points often only tell what they have found. But how they progressed to that end point is not so well expressed. And yet it is the doing of the mathematics in the long run that is the more important issue. It is not only the skills to be learnt and honed that is crucial: The general ability of how to go about solving a problem and the joy that can come from being involved in such a process is what will have more long lasting importance for students. And clearly there may be times when the mathematical values that were part of the doing, can be named.

ALAN: I do want to comment on something else. My sense is that the science people know much more about group projects and investigations. I always liked it when I was teaching at school; we had science practical and science theory: theory was in the morning and afternoon was science practical. And the practical always related to the theory and I wonder whether you could do the same with maths. You could have maths practical and you could have maths theory. Maybe we would need to change those words and have projects. I think that partly one of the difficulties for teachers to take on board some of these ideas is it is expecting them to work out how to do it. You know 'Suppose we want to do all this? Would we want to have projects at a certain time of the day?' … I took over the timetabling for Education at Cambridge. First thing I did was to block out all of Wednesday for Math Method. It was terrific. Suddenly you had a whole day to devote to a whole range of projects: Which we did. So I wonder why couldn't we do something similar in schools.
PHIL: In some ways I think you have been envisioning this for secondary schools. I reckon this would be much easier done in primary schools. You know they are far more flexible with time in that it is one teacher or a group of 2–3 teachers, that have got that group of children to teach. And they've got their space. They can tend to be quite a bit more flexible than their secondary colleagues with that and with how they organize their time as well. So it might well be a possibility there.
ALAN: It would be nice to know from some teachers … I'd like to do some case studies where teachers are trying to do new stuff with this flexible approach of using investigations and projects and doing something about values.

The Interplay of Confidence, Competence and Values

We want to have students on the cutting edge of their knowing: They need to roam their unknown. Teachers need to expect that students will indeed roam and move beyond their edge. Sadly many teachers tend to stick to one way of presenting problems that they find comfortable: a solving strategy that works for them. Another possibility is to use a variety of solution strategies for a problem and then give stu-

Starter 2: There are many teachers in primary schools who are not confident and some not competent in the mathematics they teach. Similarly, teachers in junior secondary years for whom mathematics is not their choice of teaching subject, but they are drafted into teaching mathematics (Clarkson 2016). But can these teachers who may not know their way around the mathematics apart from at an instrumental level, also work through a meta analysis of what they are doing to allow the valuing they will indulge in to emerge?

Fig. Starter 2 for our discussion

dents reflection time to discern the differences between the strategies. And within those discussions, encourage students to understand what values are embedded in their reflections.

Alan and I have been involved in university primary and secondary pre-service education programs. In one study we found that the notion of values and mathematics was nowhere apparent in such programs (Clarkson et al. 2010). Hence, part of our on-going conversation dealt with the education of teachers (Fig.).

Alan had written three points regarding this Starter:

What kinds of teachers do we need teaching mathematics? Perhaps looking at Finland might be a useful example? What support do our teachers need?

And then added another three that dealt more with us as researchers:

What do teachers currently do? What are teachers normally like? Are there examples that help us understand where are the gaps in our knowledge?

PHIL: Well one of the things about teachers choosing values … is how much mathematics do they need to know. When I wrote 'Starter 2' I was thinking about primary pre-service teacher students. Clearly there are some that come to university knowing their mathematics. They are good at maths. But there are many more … ah, well, their understanding is a bit 'iffy'. And then there is a small group of students that you really have to work with on what they do understand maths to be, and math concepts, let alone skills etc. And it did prompt in my mind, 'can they appreciate mathematical values if they are battling with just what mathematics is?' Thinking for example about say mystery as one of the values; with little grandkids, well mystery is sort of just natural for them. But is it natural for the teachers? I suspect it is not. And I think if you said 'mystery' then 'mathematics', they'd say 'What?' They would not get the connection in the way we see it. I suspect because they haven't done mathematics … It is not that they have not done enough mathematics. I suspect they have not done mathematics in a way that exposes them to these possibilities of mathematical values.

ALAN: I think you are right. But it is not just a matter of the teachers not doing what they should be doing, or at least what we think they should be doing … I'd like to think a bit more about the pressure on teachers and well how this links to 'choice'.

PHIL: Yep

ALAN: Who has the goodies that's going to be stimulating the teachers and make them brave enough to take on the challenge if you like? So, yes I think that choices are there. But I think they're (the possibility of making choices) probably for most teachers, hidden. My experience of in-service work is teachers saying 'Well that's all very well but I've gotta do the plan, I've gotta do this, I've gotta do that.' I felt this keenly when I was doing this study for ACER with Lawrence (Ingvarson et al. 2004). We were looking at different structures for (school subject) departments in secondary schools. I was quite influenced by thinking about how to characterize departments. I chose various words that to me described what a particular department was about. Maybe this happens more in primary schools, I'm not sure, but some of the ideas we toyed with then I thought were very good and the notion that you could have a maths department, with a head of department, this is very (well was) very strong in the UK (I guess I will soon see what it is like now in the UK) and strong ideas of the group notion to be important so that the teachers don't feel alone, and are made aware of the choices that are open to them and they are party to discussions.

PHIL: Yes that notion of 'groupness' in primary schools in Australia: the early years P-2 teachers often work as a group: as do the years 3–4 teachers, and the 5–6 teachers. You know the year 3–4 teachers for example work together as a well-knit group on planning, etc. So there are avenues there as well. But it is not like the mathematics group (or department in a secondary school). It is the group of year levels teachers. It is a different structure of the school. One of the real difficulties in primary is to have a teacher who is recognized in the school as the leader of mathematics. Invariably they have a Literacy leader, but for the next by far biggest block of teaching time, mathematics, most times there is no-one designated as the leader. That's totally surprising to me, but it is a rare thing to have.

ALAN: Yes it brings up again the issues that surround teachers in terms of curriculum choices and methodological choices that they have to face (Seah et al. 2016). And this is where as you were saying the choices may be recognized but they are not appreciated in the way that maximizes the potential of the subject as a value-laden subject.

PHIL: I gave that talk to teachers up in North Queensland this year (Clarkson 2017). It was more of a workshop rather than a keynote lecture that they asked for. We started with content; what are the 'big ideas' of mathematics and so on. And then halfway I inserted the notion of values and you could see quizzical looks going round the room.

ALAN: I bet you did.

PHIL: But I think from the feedback of various teachers during the remainder of the day, the notion of values was recognized as being part and parcel of the subject but, in one teacher's words, 'I've never activated it. I've never activated that part of the subject. It's real food for thought.' But the notion that it is part and parcel of this subject area I think that was something that many of the teachers present actually started to recognize. Some of them for the first time, others knew it, but no action.

ALAN: Yes that's the question: Why have they done nothing about it? You know I tend to fall back I'm afraid onto the defensive argument of, 'life's tough and ah' …

PHIL: And it is so. Certainly is for teachers!

ALAN: You can't get away from the time pressure. You actually do need to get some sensible, serious, good mathematical work done. Yes it is a difficult thing.

PHIL: But doesn't that also go back to what we were trying to talk about in that paper we wrote with Annica and Wee Tiong (Seah et al. 2016), gf that it is also the way you conceptualize the curriculum and it is also the way you conceptualize mathematics and what's important about it. If you re-conceptualize it, think about it in a different way, then it becomes a notion that 'you don't have to teach more. That this (values) is not an 'add on' and you've gotta make time for it.' It becomes more of 'you've got to teach differently'.
ALAN: Yes … what's the argument for doing that? Why do I have to teach differently?
PHIL: The Bishop told you to! Sorry. Been at the Catholic University too long!
ALAN: It seemed like a good idea at the time!!
PHIL: Well … one of the reasons you should is that you are actually getting down to the fundamentals of what mathematics is when talking about these values.
ALAN: Yep
PHIL: It is one of the reasons. It is not the only reason by any means. But is one of the reasons that actually gives it sense. Now when you talk about weaving you put the thread through the basic framework made up of the tense warp; you know those strands are the basic stationary threads that run this way. And then you put the weft through it in the pattern that you want on the fabric. But unless you know that those basic structure of threads, the warp, are there through which the weft has to go, you end with nothing. It is the warp that holds the whole fabric together: And so with the values imbedded in mathematics. No wonder so many of our school teaching colleagues think mathematics is very 'bitsy'. You do a bit of this and a bit of that and a bit more over there; and that's mathematics.
ALAN: Yes mm that's good.

This part of the conversation made me think again of issues which we had raised in the VAMP project, but still need thinking about; What stories do teachers tell regarding critical incidents in their teaching? What impact do they see of values in these moments? Do they see valuing as part of the mix in the decision process at the time, post incident, a long time after the critical incident? (Clarkson 2008a).

Mystery

I have written before on one of the six mathematical values that Alan listed, *rationalism* (Clarkson 2004). However, one that to me is undervalued is *mystery*. Many people either regard mathematics as mysterious because they do not understand it, or dislike it and hence do not want to understand it (Andersson and Wagner 2018). And yet as Alan suggested in his book it is the sense of *mystery* and its counterpart *openness* that bring understanding to the bigger picture of how mathematics sits within our broader culture. These two values go beyond the symbolic component (*rationalism/objectism*), which allows students to grapple with mathematical ideas "we think are worth knowing about", and the societal component (*control/progress*), which "shows how ideas are used" (Bishop 1988, p. 114). *Mystery* and *openness*

Starter 3: I wonder whether little kids have more of an understanding of the mystery of mathematics and that this seeps away as they age? I am pretty sure there are many students who do not get that rationalism is part and part of doing mathematics.

Fig. Starter 3 for our discussion

allow students to reflect on mathematics as a whole. "Valuing mystery … (can lead to) thinking about the origins and nature of knowledge and the creative process, as well as abstractness and dehumanized nature of mathematical knowledge" (Bishop 2016, p. 50).

Others too have thought that mystery is important in capturing students' (and teachers') interest through their imagination. This leads to a deeper appreciation of just what mathematics is. Mason (2015) notes the delight, surprise and curiosity that he was trying to invoke, and did, in his teacher audience as he involved them in various activities; surely all aspects of mystery in a good sense. Ernest (2015) in defining the beauty of mathematics includes surprise, ingenuity and cleverness, which seem to me to also speak of the mystery of mathematics that a student might (should) be experiencing.

ALAN: Someone, a scientist, said to me, 'Why have you put mystery in? Mathematicians are not terribly interested in mystery. Whereas in science, that is our bread and butter.' And that made me think … well that is, maybe the case.
PHIL: I am now playing with two grand children who are three and four. They do ask why questions, and they do get interested, and to me they sort of are really exploring at a cutting edge for them, and it is all engaging and it is a bit of a mystery for them (Fig.). 'Look what I have … Grandpa look what I have made.' 'Well of course you have made that kid, it's gotta be that way because …' I think but do not say.
ALAN: Because 'that's the way it works'.
PHIL: '… that's the way it works'. But they don't see the pattern and the obvious eventuality of if you have square blocks then you've gotta have those smooth sides … then it will end up that way. But they see it as a mystery: 'Gee look what I made! It's a mystery. How did that happen?'
ALAN: Yer yer
PHIL: I wonder whether there is something about it that we don't evoke mystery at all in our teaching of mathematics. It seems for many (most?) students if you're not sure where your work for this mathematics problem is going, then you've gotta be wrong.
ALAN: Mmmm Yes
PHIL: You sort of gotta know the end product. You can't just go and explore.
ALAN: That's right.
PHIL: I think, I reckon that it is pretty sad.
ALAN: There was some discussion after the film *The Man Who Knew Infinity* about the Indian mathematician Srinivasa Ramanujan came out. I had some interesting conversations about that. I was trying to explain to some other colleagues that he was very good at making these conceptual leaps (Pressman 2016). In the film, one of the

Cambridge mathematicians was saying "How do we know that's the case? You've got to justify this. You just can't ... just can't come up with these ideas and keep going. You've got to be able to prove. You've got to be able to substantiate this.' And yer, that was quite an interesting sequence I think. It seemed to me one could bring a little bit more of that into this conversation a bit more of that idea.

Actually in the film I thought there were two points of mystery. One was certainly the one Alan noted. But another was that Ramanujan just seemed to accept the mystery of his insights (which very occasionally did not turn out to be provable). The traditional Cambridge mathematicians could not accept the 'leaps of faith' Ramanujan made and it was a mystery to them that he accepted his leaps without question. The intervention of Hardy, facilitating the communication between the two groups, was in itself fascinating.

Students' Competence, Choice and Values

Our discussion had focused on teachers and not so much students simply because of the pressure of time. Hence our discussion of 'Starters 4 and 5' was limited (Fig.).

When thinking about 'Starter 4' I recalled that during a teacher professional learning session I summed up one point with 'Never interrupt kids who are talking maths. If students are talking mathematics, then as the teacher you may gain some insights into the thinking that is going on, but equally on reflection you may understand more of the valuing that they are choosing in that context.' We know that there is always a huge range of knowledge within a class group (Clarkson 1980). And the same is probably true for the valuing that students are choosing at any one time. But as a teacher both are important to plumb.

PHIL: I wonder too whether you had any more thoughts about the MWB construct we built some time ago now (Clarkson 2010). And whether that is a useful thing for teachers? It really has not taken off with colleagues.
ALAN: No. It's a pity that it hasn't because I think it focused quite clearly on what teachers were about and what they find rather difficult. It could still be useful I think.

Starter 4: How much understanding of mathematics, and/or doing mathematics, do students have to have before they can start understanding the roles mathematical values play? That's not year level dependent. It's the language we used in the Mathematical Well Being[2] (MWB) construct. Is there some sort of threshold of being able to do, and know you can do, mathematics before you can move to a meta-analysis state and sort out something about your values?
Starter 5: Choice is a crucial aspect of valuing.
a. Do students recognize what valuing is? What age does this kick in? Do they recognize that this behaviour is also associated with doing mathematics?
b. Students will have been making choices ever since they started doing mathematics.

Fig. Starters 4 and 5 for our discussion

Maybe keeping the two areas of mathematics and values in mind is going to push the MWB a little bit: I think that'll be useful. So I think we have a potential dichotomy, for teachers perhaps, but not necessarily in a problematic way.
PHIL: Well this dichotomy: The MWB was always trying to build a bridge between what has been set up as a dichotomy of content and values. But if you think about the different MWB stages, both were always represented in each stage: The doing mathematics, talking about it and being confident in explaining, etc. BUT the values are there as well. How could it be otherwise? They are part and parcel of the maths. That needed to be appreciated. So the two are really one, but aspects of the one.

Alan did write in his notes under 'Starter 5':

What guides choices in the classroom? Education is all about choices?

Teachers make choices before lessons, during lessons, post lessons and with regard to the holistic context within which specific lessons are located (curriculum, assessment, resources, etc.). There are also choices students can make, but they are normally within a classroom context and hence students are often constrained more in what they can change. So what choices can students make in the classroom? What are their options and what are the constraints? Indeed what allowances do we make for students to express values, compare values, and indeed think about values?

Interestingly, students can choose to disengage in various ways even within the mathematical classroom context, which is rarely a choice for teachers. Even if students stay engaged they may well choose not to reveal their value choices and at times disguise their choices for a variety of reasons. Students are schooled at a young age to know that to reveal what they really value may not be acceptable to teacher or peers, so they may keep quiet or pretend otherwise. If this assertion is correct, this calls into question whether students' actions are a good indication of their values. For students, what they are allowed, or think they are allowed to do, may well override a chosen or intended value. So maybe classrooms are not contexts that are conducive for students to reveal deep value choices. What would we discover about students' mathematical valuing if we talked and observed students doing mathematics outside of the classroom? Almost certainly some mathematical values are learnt at home before schooling begins, and some values may well continue to be reinforced by home, even when they are at variance with what teachers espouse. How do we research that issue?

In the VAMP project we soon realized that language was an issue that we needed to address when working with teachers. Not surprisingly the more we talked with the same teachers about mathematical values, a shared understanding of key ideas emerged, and a shared language which enabled us to think more deeply together. Not surprisingly a similar situation arose when working with students (Atweh et al. 2010).

An assumption that has been at the heart of our work, a good one we believe, is that teachers do have some influence on their students' values. However the reverse question might also be worth exploring: Do the values of students influence teachers' values? A further worthwhile question might be; Does the teacher's influence over curriculum, resources, assessment and teaching ethos have an impact on students' values?

If we are to make progress we might need to rethink what are good strategies for collecting data since there are so many constraints in play at any one time. Maybe we need to plan for lengthy periods of on-going data collection in differing contexts. Who asks the questions that drive the data collection should also be debated. Why should that always be the researchers? What questions would students ask? What questions would teachers ask? Why these?

Final Comments

ALAN: We have talked a lot about mathematical values, but there are the other values
PHIL: Pedagogical
ALAN: Yes and cultural values.
PHIL: It's interesting isn't it that we try to talk about mathematical values and yet being teachers we do tend to slip across to pedagogical and cultural values among others. I think that speaks to me of trying to compartmentalize these ideas. But to actually think about them in the real world of teaching there's a free flow between them. In the actual act of teaching you can't compartmentalize them. But coming back to the six values you wrote about all that long time ago, are there others?
ALAN: That's always the question.
PHIL: I think we have talked about this a number of times.
ALAN: I can't really answer that question until, until I feel comfortable with what the six are about. If you are asking me about value 7 or 8 I'd have to say, 'Hold it'; because I'm not going to give up those six lightly. I think they do strike a chord with people.
PHIL: And capture most of what you see as mathematics.
ALAN: Yer yes. I don't think that the technology has made them all irrelevant in some way. Yep I'll stick with the six for now.

Summary

This conversation has led me to ponder again the six values that Alan had articulated. In some ways the first of the three pairs (rationalism, control, openness) are most commonly acknowledged, although of these three, there seems to be more emphasis on the first and less on the third. I had originally thought of the six as somewhat discrete but now I realized my position had changed. They are each distinct but the boundaries between them are nevertheless somewhat fuzzy. For example, the description given by Bishop about *progress* and *openness* seem to allow these two to slot rather nicely together. Further the language one needs to express *progress* and *openness* overlap in particular; you do need to use logical connectives if you are making generalizations or justifying. But then they are also needed for rationalism too. Thus the overlap is an issue that I suspect needs to be explored further.

I had wondered how *rationalism* and *mystery* could coalesce. But just as Bishop suggests although he still wonders about the mystery of Pythagorean triples, among other things, he clearly knows the rational mathematics of the triples. It seems to me that at times rational understanding seems to deepen mystery, not negate it. And yet so much teaching emphasizes only the rational. How can the emphasis on the rational continue, as it should, and yet allow elements of mystery to seep in too?

The issue of whether there is some threshold of knowledge before mathematical values can be appreciated in depth still for me stands as a crucial issue. The notion of students' (and indeed teachers') choices also remains an issue that needs detailed exploration. Choice is a foundational notion when considering valuing. But how can this be undertaken in the complicated context of a classroom?

This conversation did not result in many concrete positions that we agreed on. But more importantly it continued to open each of us to further and crucial notions to explore. One hopes the reader will also be challenged to think broadly on the notion of mathematical values, a crucial element of the foundational frame that holds what we understand as western mathematics.

References

Andersson, A., & Wagner, D. (2018). Remythologizing mystery in mathematics: Teaching for open landscapes versus concealment. *Education Sciences, 18.* https://doi.org/10.3390/educsci8020041

Atweh, B., Clarkson, P., & Seah, W. T. (2010). What values do middle school students attribute to studying mathematics: A pilot study. In W. Chang, D. Fisher, C. Lin, & R. Koul (Eds.), *Envisioning the future* (pp. 9–20). Hualien, Taiwan: Department of Life Science National Taiwan Normal University, and SMEC, Curtin University.

Bishop, A. J. (1988). *Mathematical enculturation: A cultural perspective on mathematics education.* Dordrecht: Kluwer Academic.

Bishop, A. J. (2016). Can values awareness help teachers and parents transition preschool learners into mathematical learning? In T. Meaney, O. Helenius, M. Johansson, T. Lange, & A. Wernberg (Eds.), *Mathematics education in the early years* (pp. 43–56). NY: Springer.

Clarkson, P. C. (1980). Interpreting test results. *Australian Mathematics Teacher, 36*(3), 14–15.

Clarkson, P. C. (2004, Dec). 'Researching the language for explanations in mathematics teaching and learning.' *Conference paper presented at AARE.*

Clarkson, P. C. (2008a). In conversation with Alan Bishop. In P. Clarkson & N. Presmeg (Eds.), *Critical issues in mathematics education: Major contributions of Alan Bishop* (pp. 13–28). NY: Springer.

Clarkson, P. C. (2008b). Values. In P. Clarkson & N. Presmeg (Eds.), *Critical issues in mathematics education: Major contributions of Alan Bishop* (pp. 229–230). NY: Springer.

Clarkson, P. C. (2017). *Mathematical wellbeing: I'd like to see that! Keynote presentation at conference, 'Positive mathematics for indigenous students' at Mount St Bernard College*. Queensland: Herberton.

Clarkson, P. C., Bishop, A., & Seah, W. T. (2010a). Mathematics education and student values: The cultivation of mathematical wellbeing. In T. Lovatt, R. Toomey, & N. Clement (Eds.), *International research handbook on values education and student wellbeing* (pp. 111–136). Dordrecht: Springer.

Clarkson, P. C., Seah, W. T., & Bishop, A. (2010b). Mathematics well-being and teacher education. In R. Toomey, T. Lovatt, N. Clement, & K. Dally (Eds.), *Taking values education to teacher education* (pp. 179–194). Terrigal, NSW: David Barlow.

Ernest, P. (2015). Mathematics and beauty. *Mathematics Teaching, 248,* 23–27.

Ingvarson, L., Beavis, A., Bishop, A., Peck, R., & Elsworth, G. (2004). *Investigation of effective mathematics teaching and learning in Australian secondary schools.* Camberwell, Vic.: Australian Council of Educational Research.

Mason, J. (2015). Being mathematical – with, and in-front-of, learners. *Mathematics Teaching, 248,* 15–20.

Mezrich, B. (2003). *Bringing down the house*. NY: Free Press.

Pressman, E. (Producer). (2016). *The man who knew infinity* [Motion Picture].

Seah, W. T., Andersson, A., Bishop, A., & Clarkson, P. (2016). What would the mathematics curriculum look like if values were the focus? *For the Learning of Mathematics, 36*(1), 14–20.

Spacey, K., Ratner, B., Brunetti, D., & DeLuca, M. (Producers). (2008). 21 [Motion Picture].

Vilson, J. (2017). Math was never neutral. *Medium.* Retrieved October 29, 2017, from https://medium.com/@thejlv/math-was-never-neutral-173b52e9bf4a.

9

Socially Open-Ended Problems for Enriching Student Learning with Mathematical Models and Social Values

Takuya Baba and Isao Shimada

Abstract Our society is increasingly dependent on technologies such as in the life sciences and information technologies. The technologies create alternatives and the choice among alternatives is guided by what the individual values. On the other hand, Bishop (1991) pointed out the danger that the general public and students understand that mathematics learning is regarded in many countries as being unreal and value-free, mainly because of the abstract nature of mathematics. This gap between social reality and students' perceptions deprives many students of a willingness and positive attitudes towards problem-solving in mathematics. This chapter proposes a new approach to dealing with social values through problem-solving. Baba (2007, 2009) has named this type of problem "socially open-ended problem" which elicits students' social values by extending a traditional open-ended approach (Shimada 1977). This chapter describes some basic ideas and discusses how the social values are treated while dealing with socially open-ended problems.

Keywords Socially open-ended problem · Mathematical models · Social values · Problem-solving

Historical Background and Research Aim

Since the 1980 declaration of the US National Council of Teachers of Mathematics (NCTM), problem-solving has occupied the core of mathematics education. Some key ideas such as meta-cognition and problem-solving strategies have been developed through research (Schoenfeld 1983). On the other hand, Lesh and Zawojewski (2004) pointed out limitations in the research on strategies and proposed Model-Eliciting Activities (MEA) as a new approach to these limitations. MEA and problem-solving

T. Baba (✉)
Hiroshima University, 1-5-1, Kagamiyama, Higashi-Hiroshima, Hiroshima 739-8529, Japan
e-mail: takuba@hiroshima-u.ac.jp

I. Shimada
Nippon Sport Science University, Setagaya, Japan
e-mail: shimadaisao@nittai.ac.jp

look very similar. MEA emphasizes developing mathematical models from phenomena. It not only places emphasis on finding the solution to the problem but also on deeply understanding the phenomenon through the mathematical model. Here, solving the problem forms part of the MEA, but it is not all of it. This is a similar idea to the mathematization cycle within the OECD definition of mathematical literacy. But they are slightly different as well. As the words suggest, a model-eliciting activity has more of a focus on the process, whereas mathematical literacy places more stress on the children's ability to interpret the world.

Before moving further, we first review historically problem-solving in the Japanese context. In Japanese mathematics education, problem-solving has been developing in a unique way. After World War II, a US education mission was dispatched to Japan in order to prescribe the education system for post-war Japan. It recommended Life-Unit Learning (LUL) based on John Dewey's philosophy. In LUL, a problem means something in our daily life. For example, weeding in the field was used as introductory material to teach multiplication of a fraction by an integer in Grade 6 (Toda 1953). Here the portion of a rice field which can be weeded in a day was given. Then children were asked to calculate how many days were necessary to finish weeding the whole rice field. The purpose of problem solving was to use the context of daily life as introductory material to teach concepts of number, quantity, and shape and to develop the ability to consider and treat the phenomenon and scientific attitudes in daily life (Ministry of Education, Science and Culture 1947). Here, daily life explicitly appeared in problems. (In this chapter we use this characteristic but called it the "sociality" of problem-solving.) The word and concept of "mathematization", which refers to the process of developing a mathematical model from the daily life context, finding the solution, and interpreting the solution in the context, was yet to be developed.

The LUL was not a continuation of the mathematics education in the pre-war period, but rather was brought from the outside, as noted above, from the US. In addition, a decline in mathematics achievement in primary school was pointed out during this period (Kubo 1951). LUL was criticized as the cause of the poor level of achievement. Instead, a new curriculum called Systematic Learning (SL) was introduced to lead students to an understanding of basic concepts and principles of number, quantity and shapes, and to develop more advanced mathematical thinking and treatment (Ministry of Education, Science and Culture 1958). It was during this time that "mathematical thinking", the pillar of mathematics education in Japan, was introduced in the objectives of the Course of Sutdy. This shift from LUL to SL had a long-term effect on mathematics education in Japan.

The open-ended approach (Shimada 1977; English translation, Becker and Shimada 1997) started as a project headed by a team of professors and school teachers, who developed and intensively employed open-ended problems, in order to evaluate this mathematical thinking. The open characteristic of the problems stimulated children to produce various solutions and to pay attention to mathematical structures embedded in those solutions. These characteristics attracted attention from many researchers and practitioners all over Japan. It later became a teaching method to develop mathematical thinking.

In Japan, due to the experience of LUL, the term "problem solving" was avoided for some time. In the 1980s, when the results of the SIMS (Second International Mathematics Survey) were released, it was pointed out that Japanese students had attained very high achievements in general but relatively low achievements in non-routine word problems (National Institute for Education Research 1991).

Coincidentally, around the same time, the NCTM declared the 1980s to be the decade for problem-solving. This situation caused the Japanese mathematics education community to discuss the interpretation of this movement, because of its previous experience, and coined the new word "Learning Through Problem Solving" (LTPS) (Ministry of Education, Science and Culture 1989) to differentiate it from the previous problem-solving in the LUL period. Naturally, it continued to emphasize higher-order mathematical thinking, through LTPS using non-routine problems rather than simply solving a problem. In this sense, mathematical thinking is the persistent hope of the Japanese mathematics education community (Ueda et al. 2015). Since mathematical thinking emphasizes mathematical structure found in various solutions to the open-ended problem, this characteristic in SL and LTPS is called the "mathematicality" of the problem-solving.

We have so far reviewed some of the major events in the history of mathematics education in post-war Japan and noted the trend from LUL to SL and then to LTPS. The first put more emphasis on the daily life context as introductory material and had a tendency to have less emphasis on mathematics. The second (SL) and its successor LTPS puts more emphasis on recognition of mathematical structures and has a tendency to take the daily life context lightly. In summary, problem-solving and the problems considered have shifted from the "sociality" of the problem solving (the daily life context) to the "mathematicality" of the problem solving (the mathematical structure). Interestingly PISA has proposed mathematical literacy and more contextual problems (OECD 2013). However, in Japan, problems have had a greater inclination towards mathematical structure and structural thinking than towards social reality until recently because of the historical background outlined above.

Socially Open-Ended Problems

Since the 2000s, international comparative surveys of education such as TIMSS and PISA have been conducted regularly. A new type of problem-solving has been proposed as a part of the PISA study. The "problem" in the PISA study is overwhelmingly long and very contextual. The context seems to provide very authentic situations to the students (Palm 2008). Like MEA, problems in PISA studies ask students not only to find the mathematical solution but also a solution strategy which involves a series of processes such as moving back and forth between the mathematical world and real world, and hence finding a mathematical solution and its interpretation in a real world context. Here the ability required is termed mathematical literacy. Its definition is given as follows:

> an individual's capacity to formulate, employ, and interpret mathematics in a variety of contexts. It includes reasoning mathematically and using mathematical concepts, procedures, facts, and tools to describe, explain, and predict phenomena. It assists individuals to recognize the role that mathematics plays in the world and to make the well-founded judgments and decisions needed by constructive, engaged and reflective citizens (OECD 2013, p. 17).

In order to characterize mathematical literacy in the Japanese context, it is important to revisit the experience of problem-solving in Japan. After LUL, the open-ended approach has become the teaching approach in Japan to enhance mathematical thinking to consider structurally various solutions and to find the commonness and invariance among them in open-ended problems (Shimada 1977; English translation, Becker and Shimada 1997). Here Nakajima (1981), who was a leader of research on mathematical thinking, stated "Mathematics is said to hold three characteristics (abstractness, logicalness and formality). These characteristics have been naturally formed as a historical pursuit based on certain values. And these values, which form a foundation of three characteristics, are pointed out .... conscience, clarity and integration" (p. 56). He explained further that two of these values, conscience and clarity, started appearing in the Course of Study (1953), and they were refined and stated in the next Course of Study (1968) by adding integration as a source of creativity. Thus for Nakajima, the mathematical values of conscience, clarity and integration, are embedded within the nature of mathematics.

On the other hand, Iida et al. (1994) discovered that moral issues or ethical values might occur in the process of researching the open-ended approach when students deal with such topics as melon division[1] and room assignment.[2] More recently, Greer (2007) has pointed out that the mathematical modelling for proportion problems prompts recognition of equity. These findings indicates division in the real world may create issues related to ethical issues.

So far, we have seen two types of values. One consists of mathematical values, which are related to mathematical structures such as clarity and integration. The other is social values such as equity and fairness. Here our interpretation is that mathematical literacy and full contextuality of PISA problems may have the potential to integrate the two types of values. This is a new dimension of problem-solving, which is different from LUL, SL and LTPS.

The world currently faces global problems such as environmental issues and poverty issues, and individuals in a society also face personal problems such as conflict management and medical considerations. Since these problems are multifaceted due to technological advancement and the globalized economy, their resolution can have various alternatives, one of which may be chosen according to values. These problems are called trans-science problems, in which science and politics interact, and cannot be solved only scientifically (Kobayashi 2007). Perhaps the perfect resolution is not possible. For example, the problem whether we should maintain nuclear

---

[1] "Melon division" provides a situation where 10 melons as a prize in a game are to be divided among three teams and the scores for each team are given.

[2] "Room assignment" provides a situation where 10 students are to be assigned to 4 rooms of different sizes during an excursion.

Table: Comparison of two types of open-ended problems (Baba 2007, p. 22)

	Mathematically open-ended problem	Socially open-ended problem
Objective	To nurture mathematical thinking	To nurture mathematical thinking and judgement based on mathematical thinking and associated social values
Problem	To allow mathematically diverse solutions	To allow mathematically diverse solutions and associated social values
Method	Discussion on mathematically diverse solutions and their generalization and symbolization	Discussion on mathematically diverse solutions and the associated social values

power plants for energy requirement in the society cannot be solved only through technological solution.

Despite these necessities of our current society, Bishop (1991) pointed out the general public and students understand that mathematics problem solving is regarded in many countries as being unreal and value-free, mainly because of the abstract nature of mathematics (Skovsmose 1994). Since it is values that have impact on students' lasting impressions of their school mathematical experiences (Clarkson et al. 2001), the lack of positive attitudes towards mathematics learning may have an impact on individual ability and thus society's ability as a whole to solve such problems as noted above.

From these considerations, we propose a pedagogical approach for dealing with problems that can be encountered in daily life and that involves some social values in their solutions. Here this type of problem is called a socially open-ended problem (Baba 2007, 2009), because it provides different mathematical models for a solution like the traditional open-ended problems and also the social values which go together with the models. Thus employing the classification of openness (Hashimoto 2007), this approach contains open process and open-end product. This prompts a new pedagogical approach because teachers have a choice in that they may use the social values appearing in the mathematical models. So far, teachers usually have avoided dealing with social values, seeing them as noise in mathematics teaching, and regarding them as unnecessary to understanding mathematics concepts (Iida et al. 1994). Our proposal, however, is that a mathematics lesson can be enriched with social values, and thus the mathematicality and sociality of problem-solving are integrated into this approach.

We will now make a comparison between the traditional open-ended approach and the proposed pedagogical approach with socially open-ended problems. Whereas the traditional open-ended approach focuses mainly on mathematical aspects of problems, the socially open-ended problems approach utilizes both mathematical solutions and their associated social values to nurture judgement based on them (see Table for a tabulated comparison).

At a school cultural festival, your class offers a game of hitting a target with three balls. If the total score is more than 13 points, you can choose three favorite gifts. If you score 10 to 12 points, you get two prizes, and if you score 3 to 9 points, you get only one prize. A first grader threw a ball three times and hit the target in the 5-point area, the 3-point area, and on the border between the 3-point and 1-point areas. How do you give the score to the student?

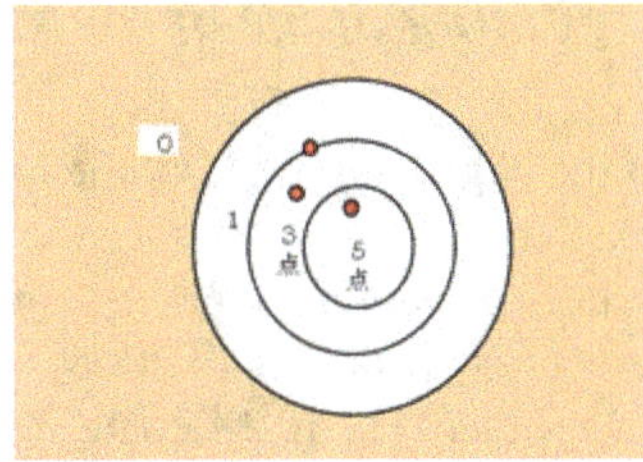

Fig. Socially open-ended problem (Matoate)

Lesson Using a Socially Open-Ended Problem

We now consider an example of a socially open-ended problem for further discussion. The authors have developed one of the socially open-ended problems, "Matoate" (hitting the target, see Fig.) (Shimada and Baba 2012, 2016). This problem was given to 38 students, comprising 19 boys and 19 girls. The second author, who taught this lesson is a teacher who specializes in mathematics education, and has 40 years of teaching experience at the time.

The lesson was carried out with fourth graders in a private elementary school in Tokyo in March 2013. Since this is a pedagogical approach, it is essential to describe not only the problem but also the teacher's interventions and interactions with students during the lesson process. In this lesson, the above problem was first presented to the students. They were then expected to work individually to find a solution to the problem and the reasons for their solution. The reason given should contain a social value(s), which was associated with their solution. After the individual work, the teacher facilitated the classroom discussion among students regarding solutions as mathematical models, and the associated social values. Sometimes such social values were not explicit at first and only became explicit after the interaction. At the end of the lesson, the students were asked again to choose mathematical models and recognize the social values in the solution process.

Through the lesson the teacher was on the look out for any students who modified their mathematical models and/or changed their choice of social values, which were called transformation of social values. If this did occur, such transformations were to be ascertained through a comparison between the values at the beginning of the lesson and those at the end.

Beginning Stage of the Lesson

At the beginning stage of the lesson, the students were given the above problem and asked to write their solutions with reasons. There were two types of values that were identified, namely, equality among the whole participants and priority to a specific person (Nagasaki et al. 2008).

(Protocol [1] of "Matoate" (hitting the target)) (T stands for a teacher and S stands for a student.)

T1:	So, please think about this problem and also write the reason
	After students worked on the problem individually for 10 min, discussion started on different ideas
S1:	The first grader might be happy to get the bigger score because the ball is between the 1-point area and the 3-point area. They can get two prizes because $5 + 3 + 3 = 11$
S2:	$5 + 3 + (3 + 1) = 12$. Since it is a first grader, both points would be given
T2:	That's a great service to give both points when the ball is on the borderline between two areas. It is very kind of you to a small child

In the dialog above, S1 made a decision to give the higher score, but S2 went further and suggested giving both scores, when the ball was on the border of two areas in the target. These answers showed the social values of the students to care for the first grader. They also developed the mathematical models, S1 giving the higher score ($5 + 3 + 3$), and S2 adding both points ($5 + 3 + 3 + 1$). As can be seen, different mathematical models can be made based on the same social value "kindness" (to the first grader).

The Development Stage of the Lesson

The lesson progressed with the teacher asking for more discussion on the various solutions that the class had developed. During this process, one social value was observed more explicitly ("kindness to a specific person"), while the other social value ("fairness and equality to the whole") stayed rather implicit (see Table).

This becomes evident in the following discussion:

T3:	So, please make a presentation on how you think about this problem. S.J., please
S.J.1:	I gave 3 points to the first grader, so I wrote $5 + (3 \times 2) = 11$, 11 points, because the first grader should be welcomed (Fig.)

(continued)

(continued)

T4:	Does anyone have any questions for S.J.?
S4:	I think that we do not have to write parentheses in the expression because we calculate the multiplication before the addition
S.J.2:	Thank you, I understand. I will rewrite it as $5 + 3 \times 2 = 11$
T5:	And who did you think of, S.J.?
S.J.3:	I thought of the first grader
T6:	I will write the words "kindness to the first grader" next to the S.J.'s idea. Next, please present your idea, K.K.
K.K.1:	The ball is on the boundary of 3 points and 1 point. I give 1 point because the 1-point area of the ball is larger than the 3-point area of the ball. So, $1 + 3 + 5 = 9$, 9 points (Fig.)
T7:	Does anyone have any questions for K.K.?
S5:	What points will you give to the first grader when the ball reaches the middle just above the line?
K.K.2:	I will give 2 points
S6:	What points will you give to the first grader when the ball reaches the middle of just above the line of 1 point and 0 point?
K.K.3:	I will give 0.5 points
T8:	S.J. gave 3 points to the first grader. And who did you think of, K.K.?
K.K.4:	I thought about all the people who play the game. I want to be impartial to all people
T9:	So I will write the words "fairness to all people" next to K.K.'s idea

In the interaction, kindness to the first grader appears rather easily, but fairness to all the participants can appear only when a comparison is made. Explicitness and implicitness of the social values are also manifested in Table. For example, for the mathematical model "a. $5 + 3 + 3$", 92.9% of students wrote reasons representing

Table: Students' mathematical models and associated social values at the beginning stage (N = 38)

Mathematical models	Associated social values	Percentage of explicit social values
a. $5 + 3 + 3$	Kindness to the first grader (specific person)	92.9 (13/14)
b. $5 + 3 + (3 + 1)$		100.0 (1/1)
c. $5 + 3 + 3 + 1 + 1$		100.0 (1/1)
d. $5 + 3 + 2$		100.0 (2/2)
e. $5 + 3 + 2$	Fairness and equality to the whole participant (all students)	0.0 (0/9)
f. $5 + 3 + 1$		0.0 (0/10)
g. $5 + 3 + 3$		0.0 (0/1)

Note In the column for the percentage of explicit social values, the fractions in parenthesis showed the number of students who wrote a particular mathematical model and expressed the social values explicitly against the total number of students who wrote a particular mathematical model

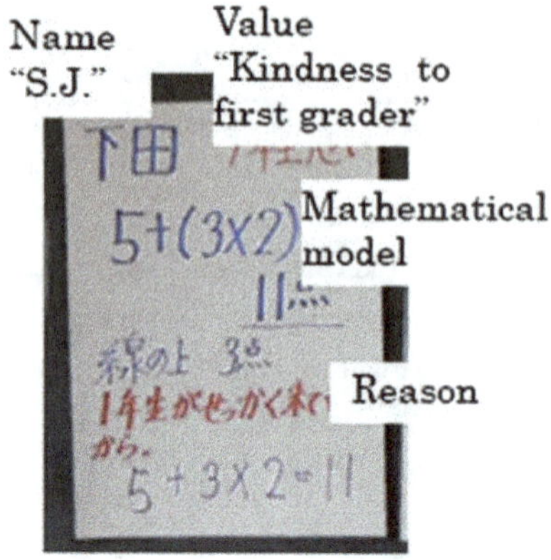

Fig. S.J.'s presentation

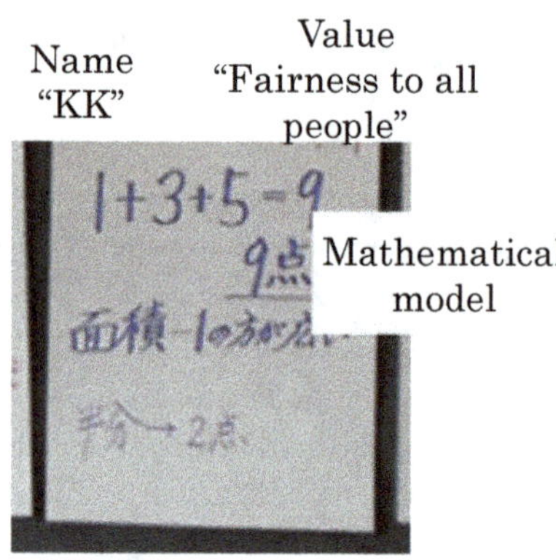

Fig. K.K.'s presentation

social values. The interaction gives an impression that they seem to be conscious of these values and even recalling the days when they were at first grade.

The reason why fairness and equality are rather latent may be that it is too obvious for the students to think about all the participants as a whole. Therefore, until some critical moment arrives, they do not pay much attention to those obvious values such as fairness and equality. In the above lesson, the students had a critical moment by being asked "who do you think of..." in comparison with "kindness to the first grader". We think that it is important to nurture understanding and appreciation of both social values in a democratic society.

The Summary Stage of the Lesson

After discussion in the classroom, students were asked to select one model and to write down the reason. Students seemed to have been influenced by the class discussion and revised their mathematical models and transformed associated social values. These are shown in Table.

Table is a cross-tabulation showing the relationship between the social values at the beginning and the summary stages of the lesson. Interestingly, the percentage of students who selected different social values at each stage, and thus transformation of mathematics is about one-third of students ($15.8 + 15.8 = 31.6\%$).

Table shows the reasons why the six students (15.8%) who selected the social value of "fairness and equality" at the beginning stage, then changed their choice at the summary stage. Half of these students (U.K., T.M. and N.M.) supported K.U.'s idea which was "$1 + 3 = 4$, $5 + 3 = 8$, $8 + 4 = 12$, $12 + 1=13$. I'll give 4 points combining 1 point and 3 points for the first grader. I give 1 point with a further bonus". Appreciating this idea, U.K. said, "The first grader will be happy and come here again." On the other hand, K.U. himself transformed his idea to the idea of T.R. which at the summary stage was given as "$5 + 3 = 8$, $(1 + 3) \div 2 = 2$, $8 + 2 = 10$. The ball is on the boundary of 3 and 1. It is 4 by adding 1 and 3, and then becomes 2 by dividing 4 by 2. It becomes 10 when I add 8 and 2". K.U. transformed his social value after knowing the other student's social value of "fairness and equality," and stated, "I think that it is nice to give 2 points because of equality."

Table: Percentage of students in class who choose associated social values at the beginning and summary stages of the lesson

		Associated values at the summary stage		
		Fairness and equality	Kindness to the first grader	Total
Associated values at the beginning stage	Fairness and equality	36.8 (14/38)	15.8 (6/38)	52.6 (20/38)
	Kindness to the first grader	15.8 (6/38)	31.6 (12/38)	47.4 (18/38)
	Total	52.6 (20/38)	47.4 (18/38)	100.0 (38/38)

Table: Students who transformed their social values and their reasons

Name	Transformation	Reason for the transformation
U.K.	To K.U.'s idea	The first grader will be happy and come here again
T.M.	To K.U.'s idea	It is good for us to give a bonus to the first grader
N.M.	To K.U.'s idea	It is good for us to be kind to the first grader
K.U.	To T.R.'s idea	I think that it is nice to give 2 points because of equality

Discussion

Three key points arise from the above results and are emphasized here. Firstly, some social values are implicit. Mathematics has historically been regarded as culture-free and value-free (Bishop 1991). Through its characteristics of abstraction and symbolization, it loses the original context and becomes free from such contexts and values. Such abstractness can be a source of power for mathematics, and thus mathematics can be applied to many different contexts. On the other hand, students in solving the above 'Matoate' problem provided various mathematical models and the social values associated with the models. Indeed students argued that the social value, which gave rise to the mathematical model, becomes a reason for the model. Nevertheless the mathematical model and social value are quite separate and should not be conflated, although they are associated entities. Although it is important for students to be conscious of this reasoning and thus social values, these data also revealed some social values were rather implicit. In order to make these implicit social values explicit, the teacher entered into quite a discussion with the students. In other words, the teacher was quite conscious of the possibility that the social values embedded in the problem may well be quite implicit for some students.

Secondly, let us consider the relationship between this pedagogical approach and mathematical literacy. It is an important task to develop a model to interpret mathematically the phenomenon represented by a problem. If we are satisfied with simply introducing daily phenomena in problem-solving, then we may fall into the same limitations as LUL. On the other hand, if we were to focus only on understanding the mathematical structure, it is just like the mathematics modernization movement. Today we live in a society and age in which we are exposed to significant change caused by technological advancement and globalization of the economy and society. On the one hand, this means that our options often increase. Whereas, on the other hand, various and sometimes opposing social values are produced. In such a context, it is important for students to know there are a variety of social values that can be elicited by a social context, and for them to be able to state their social value(s) associated with a mathematical model and to understand that others' may well choose different models and their associated social values. Using socially open-ended problem offers such an opportunity, where students are placed in a context to develop mathematical models associated with their social values, and also to discuss the possibility of different models and social values. This is precisely the sort of competence being promoted through mathematical literacy.

Thirdly, it is worth considering the transformation of mathematical models and social values through discussion. In the example taken here, refinement of students' thinking regarding mathematical models were observed by comparing those at the beginning and the summary stages of the lesson. On the other hand, value transformation is a little more complicated. We observed that about one third (31.6%) of students had been influenced by other students' mathematical models and the associated social values (see Table), and thus selected different social values at the beginning and the summary stages. In other words, two thirds have not changed their values. We think that even among those who have selected the same social values, some may have transformed a little within the same category, by adjusting the mathematical model and/or social value. It may have also been that some students who did not change their choice became more convinced of their original position, although we have no data that speaks to that possibility. Besides, both social values "kindness to the specific person" and "fairness to the whole group" are equally important, and we actually see their application in our daily life. In fact, social welfare should adopt both aspects at a certain level. Therefore, this pedagogical approach requires that the students will be exposed to multiple social values for a long period, understand the difficulty and importance of the coexistence of different social values, and thus discuss the models and social values deliberately, logically, and critically. It is in this context that a new value, which puts emphasis on coexistence of different values, will be created at the meta-level. A future issue is to engage with the long-term transformation of values and development of practices in secondary education through socially open-ended problems.

References

Baba, T. (2007). Primary mathematics education in a society and times with multiple values. *Journal of Japan Society of Mathematics Education, 89*(10), 20–27. (In Japanese).

Baba, T. (2009). Analysis of socially open-ended problems in mathematics education from the perspective of values. *Journal of Japan Academic Society of Mathematics Education Research in Mathematics Education, 16,* 247–252. (In Japanese).

Becker, J. P., & Shimada, S. (Ed.). (Translated in 1997). *The open-ended approach in mathematics education: A new proposal for teaching mathematics.* Reston: National Council of Teachers of Mathematics.

Bishop, A. (1991). *Mathematical enculturation: A cultural perspective on mathematics education.* Dordrecht: Kluwer Academic Publishers.

Clarkson, P. C., Bishop, A., Seah, W. T., & FitzSimons, G. (2001). An undervalued legacy of mathematics education. In G. FitzSimons, J. O'Donoghue, & D. Coben (Ed.), *Adult and lifelong education in mathematics* (pp. 47–70). Melbourne: Language Australia.

Greer, B. (2007). Sense of proportion for social justice. *Philosophy of Mathematics Education Journal, 21.* http://people.exeter.ac.uk/PErnest/pome21/.

Hashimoto, Y. (2007). Eliciting mathematical ideas from students: Towards its realization in Japanese curricula. In H. Fujita, Y. Hashimoto, B. R. Hodgson, P. Y. Lee, S. Lerman, & T. Sawada (Eds.), *Proceedings of the Ninth International Congress on Mathematical Education* (pp. 133–134). Dordrecht: Kluwer Academics Publisher.

Iida, S., Yamashita, A., Sumi, M., & Komori, A. (1994). Research on recognition of values in open end problem in mathematics education (1): Summary and preliminary report. *Research Journal on Kyushu Mathematics Education, 1,* 32–43. (In Japanese).

Kobayashi, T. (2007). *Era of trans-science: Connecting science and technology with society.* Tokyo: NTT Press. (In Japanese).

Kubo, S. (1951). *Academic achievement and intelligence test: Higher grades at elementary school in Yokohama city in Showa 24.* Tokyo: Tokyo University Cooperative Press. (In Japanese).

Lesh, R., & Zawojewski, J. (2004). Problem solving and modelling. In F. K. Lester (Ed.), *Handbook of research on mathematics teaching and learning* (2nd ed., pp. 763–804). Charlotte, NC: Information Age Publishing.

Ministry of Education, Science and Culture. (1947). *Course of study for primary and secondary school (Draft)*. Tokyo: Ministry of Education, Science and Culture.

Ministry of Education, Science and Culture. (1958). *Course of study for primary and secondary school*. Tokyo: Ministry of Education, Science and Culture.

Ministry of Education, Science and Culture. (1989). *Course of study for junior secondary school*. Tokyo: Ministry of Education, Science and Culture.

Nakajima, K. (1981). *Mathematical thinking at primary and secondary mathematics education: Consideration of its development*. Tokyo: Kaneko Shobo. (In Japanese).

National Institute for Education Research (NIER). (1991). *International comparison of mathematics education—The final report of second international mathematics study*. Tokyo: Dai-ichi-Houki. (In Japanese).

Nagasaki, E. Shimada, I., & Nishimura, K. (2008). Study on the refinement of the structure of the power to connect mathematics and society. In *Proceedings of the 41st Annual Conference of Japan Society of Mathematics Education* (pp. 231–236). Tokyo: Japan Society of Mathematics Education. (In Japanese).

OECD. (2013). *PISA 2012 assessment and analytical framework: Mathematics, reading, science, problem solving and financial literacy*. Paris: OECD Publishing.

Palm, T. (2008). Impact of authenticity on sense making in word problem solving. *Educational Studies in Mathematics, 67,* 37–58.

Schoenfeld, A. (1983). Beyond the purely cognitive: Belief systems, social cognitions, and metacognitions as driving forces in intellectual performance. *Cognitive Science, 7,* 329–363.

Shimada, S. (Ed.). (original version 1977, new version 1995). *Open-ended approach in mathematics*. Tokyo: Toyo-kan shuppan. (In Japanese).

Shimada, I., & Baba, T. (2012). Emergence of students' values in the process of solving the socially open-ended problem. In T. Y. Tso (Ed.), *Proceedings of the 36th Conference of the International Group for the Psychology of Mathematics Education* (Vol. 4, pp. 75–82). Taipei, Taiwan: PME.

Shimada, I., & Baba, T. (2016). Transformation of students' values in the process of solving socially open-ended problems: Focusing on long-term transformation. In C. Csikos, A. Rausch, & J. Szitanyi (Eds.), *Proceedings of the 40th Conference of the International Group for the Psychology of Mathematics Education* (Vol. 4, pp. 187–194). Szeged, Hungary: PME.

Skovsmose, O. (1994). *Towards a philosophy of critical mathematics education*. Dordrecht: Kluwer Academic Publishers.

Toda, K. (Ed.). (1953). *Learning mathematics grade 6*. Tokyou: Gakkou-Tosho.

Ueda, A., Baba, T., & Matsuura, T. (2015). Values in Japanese mathematics education from the perspective of open-ended approach. *Teaching Innovations, 27*(3), 69–82.

10

Values of the Japanese Mathematics Teacher Community

Douglas Lyman Corey and Hiroyuki Ninomiya

Abstract In this chapter, we analyse three fundamental practices of Japanese mathematics teachers to better understand the set of community values that influence their work as teachers. The three fundamental practices are: writing detailed lesson plans, kyozaikenkyu (a planning practice), and emphasizing student mathematical reasoning in instruction. An analysis of these community practices resulted in eight potential values of the Japanese mathematics teacher community. These values help the field better understand why Japanese teachers engage in the work of teaching the way that they do.

Keywords Japanese mathematics education · Japanese mathematics teaching values · Japanese in-service training

Introduction

Early work by Alan Bishop on values and valuing, and the closely tied work to cultural aspects of teaching and learning, focused on the mathematics classroom (Bishop 1988). That is, the values studied were those of western mathematics portrayed, implicitly or explicitly, by the teacher to the students. Researchers have extended the focus to include the values of the teacher and the values of the students (Seah and Wong 2012; Law et al. 2012). In this chapter, we continue to extend the scope of the values and valuing research. We extend it beyond individual teachers to a particular teaching community, Japanese mathematics teachers. This community level view in many ways hearkens back to Alan Bishop's original work on the six values of western mathematics (rationalism, objectism, control, progress, openness, mystery), since those values are really about the values of a community, mathematicians working in

D. L. Corey (✉)
Brigham Young University, Provo, UT 84602, USA
e-mail: corey@mathed.byu.edu

H. Ninomiya
Saitama University, Saitama, Japan
e-mail: hiro2001@mail.saitama-u.ac.jp

the western tradition. In this study, we look at three particular community practices of Japanese teachers and analyze these practices to uncover potential values of the Japanese mathematics teacher community. We also build on already existing work about values in Japanese culture and hypothesize how those get embodied in the community of Japanese mathematics teachers.

Framework

As a field, we have not yet built a consensus on what should (or should not) be considered a value (Bishop et al. 2003). Like many other constructs, there remain to be several different definitions of values (in mathematics education). This has also been complicated by other considerations, such as the unit consideration (personal or community/cultural values), the distinction consideration (How are values different from other related constructs such as beliefs, emotions, preferences, and orientations?), and the uncovering consideration (How do we find out what values individuals and/or communities hold?). Some previous work has been done on these issues (see the review by Bishop et al. 2003) but there is more work to do (Cai and Garber 2012). In this section, we explain how we view values at the community level in contrast to the individual level, since community values are the focus of our study.

A general definition of a value has been drawn from the writings of many scholars by Bishop et al. (2003). (We refer the interested reader to their article for the long list of scholars that they drew upon for their definition). These scholars define a value as "a construct or ideal, which refers to the desirability, preference, worthiness, priority, moral rightness, or the potential benefit of particular objects, phenomena, actions or goals" (p. 723). A powerful distinction between beliefs and values can be made by thinking about beliefs attached to the dichotomies 'true/false' or 'correct/incorrect', while values are connected to the dichotomy of 'good/bad' (Kluckhohn 1962) or 'desirable/undesirable' (Rokeach 1973).

A statement of a value always involves two aspects that often remain implied: first, the individual or community holding the value and second the object, idea, or behavior that is the phenomenon of valuation. Bishop (1988) listed six values of western mathematics. Implied in this statement is the community (mathematicians working in the western tradition), and the phenomenon (the practice of mathematics). Of course this strict distinction is an ideal because when trying to understand the valuing of cultural phenomenon, it may be hard to separate the phenomenon and the value of that phenomenon (remove the value and that could significantly change the meaning and nature of the phenomenon).

Figure suggests some relationships between behaviors and values and between individuals and communities. Of course, this is an oversimplified model and other influences on behavior exist beyond what we highlight with this diagram.

We have tried to highlight three different levels where values may be present: The top-level culture (in our case a country), the community (in our case, Japanese mathematics teachers), and the individual (a particular mathematics teacher). In reality, an

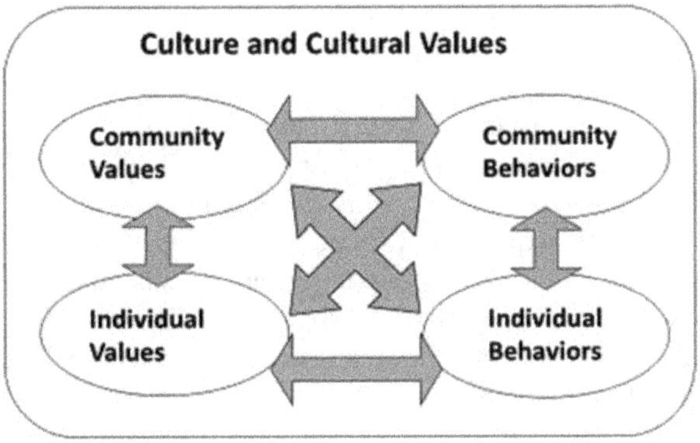

Fig. The influence of behaviors and values at the individual and community level

individual is part of many different communities and may be influenced by multiple top-level cultures. An individual's values would be influenced by the values of all of those communities and cultures and their past experience.

We do not imply with our arrows that values determine behavior, nor that behaviors of individuals determine community behaviors or values, nevertheless, although not a determinative influence, the arrows indicate that there is an influence. Values can influence behavior and behaviors can influence values, since values are not fixed but can evolve overtime, and likewise behaviors may also change over time in response to external pressures apart from values.

Community behaviors are not the same as the aggregate of individual behaviors (although that is an important part), since communities can engage in behavior or develop institutions that are beyond aggregate behaviors of individuals. Although the arrows are drawn as the same size (representing the same strength), depending on the community that is the focus of study, some arrows could carry less influence and could be drawn thinner or dashed (depicting less influence), but that would depend on particular communities and the relationships of individuals within the community.

It is also useful to discuss briefly some top-level cultural values for the subjects in our study, being Japanese, because these top-level cultural values may be significant in helping us understand values of the Japanese mathematics teaching community. There are Japanese ideas that are very difficult to describe in English, or cannot be captured adequately through simple translation, since no word or phrase is sufficient. Moreover, these ideas are fundamental to being Japanese or understanding Japanese culture. Six such ideas were discussed by Wierzbicka (1991): *amae, enryo, on, wa, giri*, and *seishin*. We cannot explain all of these here in detail, but the last two may add insight to our study. Of course, our brief explanation of these will be insufficient but such an explanation will be better than nothing. *Giri* is the feeling of obligation one should feel when someone does something nice to you or to your employer. *Seishin* is the importance of working hard toward a worthwhile goal, even if it will take a very long time to achieve. These two characteristics could place a heavy burden on teachers, or any worker, to work hard at their craft to become the best teacher they can and serve their students (which in one sense are the employers of the teacher) to the best of their ability.

Two other notions of Japanese culture that are hypothesized to have connection to mathematics teaching by Baba et al. (2012) are the ideas of *jutsu* and *waza*. "This *jutsu* concept involves aiming to pursue its object to the fullest extent, and in the process to acquire the very nature of its technique, called the *waza*… the pursuit of *waza* goes beyond simple technical aspects, and leads to nourishment of the spirit and personality formation" (pp. 31–32). Baba et al. (2012) explains that it appears from their analysis that mathematics teachers in Japan view their work as a *jutsu*, and they can acquire the *waza* of their craft through extensive study, practice, and hard work.

Current Study

Context of the Overarching Study

This chapter presents the results that have come from a larger study. A group of three US and three Japanese mathematics education researchers has been engaged in a long-term cross-cultural study focused on understanding the nature of high-quality instruction as well as the work of teaching in the two countries. Our study has been ongoing from 2011. As part of that study we have been engaged in watching public-school mathematics instruction (from 2nd to 10th grade) together in both countries (mainly focused in the greater Tokyo metro area in Japan and the intermountain west in the U.S.). These include in person observations as well as video recordings. We have also observed mathematics education courses and professional development, particularly lesson study in Japan. Since the overarching purpose of the larger study was to better understand the nature of high-quality instruction, the sample of teachers in the study were teachers that researchers felt were remarkable at teaching mathematics.

Data collected as part of this larger study include 10 videos of lessons in Japan, 6 videos of US lessons, 4 videos of Japanese post-lesson discussions as part of Lesson study, and videos of conversations between the Japanese and US researchers discussing their observations of lessons (about 20 hours). All Japanese lessons and discussions were translated into English. Research notes were taken for all of our meetings, particularly meetings via SKYPE, which were not recorded. Some semi-structured interviews were used to better understand a particular practice in Japan, *kyozaikenkyu* (see Mellville 2017, for interview details).

We have engaged in ethnographic-type participant-observer experiences in each other's countries to better understand the phenomenon of our study. About twice a month, we engage in regular conversations via SKYPE to continue our collaboration when we are not meeting in person. We have also interviewed teachers from both countries about their teaching and their teacher planning practices and their professional development activities.

Analysis

For the study described in this chapter we wanted to better understand the values of the Japanese mathematics teacher community. We performed two different analyses in this study, a thematic analysis to generate a list of potential values of the Japanese mathematics teacher community, and a member checking analysis to test the agreement of a large group of Japanese teachers with our generated list of potential values. We cover each in turn, but first discuss one important methodological difficulty in discovering community values based on community behaviors.

Uncovering possible community values contributing to a particular behavior poses a problem. An attempt to ask members of the community (Japanese mathematics teachers) about why they engage in the particular behavior (*kyozaikenkyu*, for example) would result in many different responses. However, it seems illogical that a constellation of different individual values leads individuals to engage in a specific, unique behavior. It seems much more likely that there is a set of deeper culturally driven values, perhaps supported by institutions and institutional environments, that has a strong influence on all of the Japanese mathematics teachers.

The issue here is that the teachers' initial responses may be another outcome of a particular value (or set of values) and not the value itself. In the same way that we could probe why they do *kyozaikenkyu*, we could probe why their initial responses are important. We could continue to ask such questions about their responses until the teachers are drawing on the most fundamental ideas to their practice, or the values that influence their practice. It still might take some methodological analysis to articulate the shared value(s) because the value(s) might still be implied by the responses. In our particular example of *kyozaikenkyu*, we might discover that all of the teachers are sharing ideas related to the great responsibility they feel for their students' learning. This sounds much more like a community value that is influencing Japanese mathematics teachers (community) to engage in *kyozaikenkyu* (a specific behavior).

Thematic Analysis

Our first analysis was a thematic analysis for latent patterns (Boyatzis 1998) in our observational data and discussions and it resulted in the listing and description of 8 community values. To develop these values we began by picking three community behaviors that emerged in our larger cross-cultural study as engaged in by the Japanese mathematics community but not widely practiced by US mathematics teachers. The behaviors are: (1) Emphasizing student mathematical reasoning and thinking in instruction, (2) *Kyozaikenkyu*, and (3) Detailed lesson-plan writing. Our study is not a comparison study between Japan and the US. We have used the comparison to the US here to find practices that are more likely to be uniquely Japanese. The idea is that the more unique the behaviors to the Japanese, the better chance we have of finding values of Japanese mathematics teachers

For each of these three community behaviors the two authors (one each from the US and Japan) generated a list of possible community values. This process began by hypothesizing possible values in back-and-forth conversations between the two authors of the paper. The role of the US researcher was largely that of an outside observer that knew a lot about teaching mathematics in Japan, but was not encultured into the Japanese teaching community. This allowed the generation of hypothesized values that may have been difficult to see by someone within the Japanese teaching culture. Hypothesized values were then initially vetted by the experienced Japanese mathematics education researcher, who could at least partially evaluate the credence of the hypothesis as a Japanese teaching cultural insider and as specialist in the field. The US researcher who used knowledge gained from the larger study to generate potential confirming or disconfirming evidence of a particular value also vetted hypothesized values. Hypothesized values were either rejected or revised and then discussed again. The revisions often came in the form of hypothesizing a deeper community value that may be more fundamental than one or more other hypothesized values. Seeing a pattern (Bolyatzis 1998), or underlying theme, across multiple potential values often produced a more fundamental hypothesized community value. This analysis resulted in eight values associated with at least one of the three community behaviors that were fundamental and passed the vetting process of both researchers. The eight potential values are: logical thinking as a life skill, deep understanding of mathematics, being true to the mathematics discipline, responsibility for student learning, adaptation to students, mastering the teaching craft, responsibility for community improvement, openness of teaching practice. These are explained in detail later in the chapter.

Member Checking Analysis

In order to triangulate our results we performed a small respondent validation study by asking Japanese teachers about their agreement to our results. A survey was administered to 84 elementary and junior high school teachers asking them to rate the extent to which the Japanese mathematics community shared these eight identified values (not their individual agreement) on a four point scale (strongly agree, agree, disagree, strongly disagree). Although asked specifically to consider the values of the Japanese mathematics teacher community, and not their own values, the teachers surveyed may have indeed considered their own values in responding to the survey. However, if there is still overwhelming support for these values, even if teachers considered their own values on the survey, we consider that as strong evidence that they are values of the teaching community. If participants disagreed, they were asked to make comments about why they disagreed.

Results

For each of the three phenomena we describe the behavior, why it is of interest to us, and what community values may be strongly influencing each. After describing each of the eight values, we share the results of the teacher survey.

Emphasizing Student Mathematical Reasoning and Thinking in Instruction: Behavior

As part of our study, both Japanese researchers and US researchers observed a US teacher's high school lesson on the topic of inverse functions. The response of the US mathematics education researchers was underwhelming. They (including the lead author of this chapter) pointed out many problematic issues with the lesson: the teacher often spent too much time with one pair of students, the teacher would let anyone respond to questions (basically whoever was most vocal and persistence) and, in this class, that meant assertive males were answering the vast majority of questions and making the most comments, and the teacher let some segments of the lesson go too long. The US observers thought it was a mediocre lesson for these reasons.

The Japanese cohort pointed to different characteristics of the lesson. The teacher had asked the students to find both compositions of two linear functions. The students did so and the results were, of course, x. The students started to wonder why this was and started to ask questions about the phenomenon. After some class discussion about functional processes the teacher then had the students write out, in words, the process certain functions applied to numbers. Then the students wrote out the process, in words, that would "undo" the process. For example, students wrote $f(x) = 3x + 2$ as the function process that "multiplies a number by three then adds two." The students noticed that the "undoing" process had the inverse operations but in the reverse order of the original function. Using the same example as before, the students realized that to undo the process represented by $f(x) = 3x + 2$ the students would need to "subtract two and divide by three," or represented differently, $g(x) = (x - 2)/3$. Consequently, there was more class discussion about this case. Finally, the class looked at graphs of two functions ($f(x) = {}^3\sqrt{(2x + 2)}$, $g(x) = 1/2\ (x^3 - 2)$), these complicated shapes, chosen carefully, helped the students consider why the graphs of a function and its inverse are reflections of each other across the graph of the line $y = x$.

In contrast, the US mathematics education researchers only looked at teacher and student interactions and some management issues, as noted above. But the Japanese, as well as seeing what the US observers saw, also looked at the mathematical reasoning and mathematical activity in which the students were engaged. Hence, the focus of the Japanese and US observers were quite different. The Japanese were very impressed that the teacher had set up a situation where students were authentically puzzled and really wanted to know what was going on when they kept getting "x" as the answer for the composition of functions. Interestingly, the Japanese observers agreed that all of the issues raised by the US teachers were valid concerns. However, the Japanese considered that mathematical reasoning and the authentic questions from the students to be significantly more important factors in the quality of the lesson, in effect, trumping the problematic issues of the lesson.

Emphasizing Student Mathematical Reasoning and Thinking in Instruction: Values

What values might lead to Japanese teachers to prize the mathematical reasoning and problem solving of students in class? Corey et al. (2010) argued that the intellectual engagement of students was the primary principle that Japanese cooperating teachers emphasized with their student teachers. However, why is it prized? Is it valued in and of itself, or might there be deeper values that influence teachers to prize the mathematical thinking and reasoning of students?

Our analysis of this particular behavior generated three potential community values.

Logical Thinking as a Life skill

Japanese teachers have been emphasizing the need for students to improve their ability to think for many years (Katagiri 2004), perhaps because of the changing skills needed for future employment, where automation are replacing many jobs. Whatever the cause there is a strong feeling that students need to improve their reasoning and mathematical problem solving skills, which is linked to the wider notion of 'thinking skills'. The Ministry of Education recently coined the phrase "Ikiru Chikara," translated as "surviving power" to capture this dimension of education that education should give students the ability to manage problems in everyday life.

Deep Understanding of Mathematics

Relational understanding (Skemp 1976) or deeper understanding of mathematics is a fundamental goal of teaching mathematics in the Japanese mathematics teacher community. To quote one teacher, "The goal of mathematics learning, for the type of students who memorize mathematical facts and adopt them to mathematics problems, will just be 'getting a higher score.' That is not the real purpose of mathematics education. We would like to help them discover mathematical ideas by themselves, for their deeper and relational understanding." As experienced teachers know, developing deep mathematical understanding is difficult to do without having students think and reason about mathematics themselves.

Being True to the Mathematics Discipline

According to Bishop (1988), one of the main values of western mathematics is *Rationalism*, the use of logical and hypothetical thinking. Japanese mathematics teachers are true to the discipline of mathematics as they emphasize logical reasoning and mathematical thinking. Other values of western mathematics listed by Bishop are also prevalent in the teaching of Japanese teachers, the most salient being *Mystery* and *Openness*. Japanese teachers emphasize *mystery* by using problems that have surprising results or allow students to find interesting patterns. *Openness* is emphasized as students work together to solve problems and as teachers conduct a whole-class discussion of selected student solutions (*neriage*) to deepen students understanding.

Kyozaikenkyu: Behavior

Many Japanese teachers engage in a practice called *kyozaikenkyu,* which is translated as "materials research". This phrase refers to a set of activities as part of lesson preparation. It is mentioned in the literature mainly as a part of lesson study (see the review by Mellville 2017), but it is also undertaken for every day lessons. *Kyozaikenkyu* is not just another name for lesson preparation, since there are some preparation activities that are generally not considered *kyozaikenkyu* (such as making copies for students or typing up a lesson plan). *Kyozaikenkyu* is a cultural phenomenon. There is no explicit definition of what is or is not *kyozaikenkyu* and it is not explicitly taught in teacher preparation programs. Teachers learn the practice from others as they engage in lesson study and interact with their colleagues.

For many teachers, the primary activity during *kyozaikenkyu* is carefully analyzing the textbook to understand the mathematics, considering student thinking about the mathematics, what might the key teaching questions be (hatsumon), and the flow of the lesson. Teachers often compare the approaches of multiple textbooks to improve their understanding and to craft the best lesson they can. Other activities teachers engage in during *kyozaikenkyu* include studying the previous year's lesson plans/lesson notes, studying published lesson plans from lesson study groups or lesson plans from colleagues, or reading books written for teachers (common in any commercial bookstore in Japan). The teachers study these materials to solidify the necessary aspects of a problem-solving lesson: a set of clear goals, the big idea (kadai), mathematical task (mondai), key questions (hatsumon), select and sequence student thinking, whole class discussion (neriage), and boardwork (bansho).

Kyozaikenkyu: Values

Certainly there are practical aspects to *kyozaikenkyu*. The primary goal of *kyozaikenkyu* is to develop a lesson plan and make instructional choices. However, there is also a puzzle here. Why do Japanese teachers spend time studying materials and creating their own lessons (*kyozaikenkyu*) when they have high quality published lessons that they could use without modification in their class? There is a phenomenon called "corridor *kyozaikenkyu*" where the teacher is preparing while walking down the hall to the classroom. This is frowned upon and these lessons are not considered good lessons. The following values might add some insight into this puzzle.

Responsibility for Student Learning

Japanese teachers feel an overwhelming obligation to help their students learn. Teachers naturally feel that it is their fault if students are not learning well, and so the teacher must redouble their efforts to help their students. We described an alternative situation to teachers in the larger study in which a teacher could justify that it was the students' responsibility to learn as long as the teacher prepared a lesson and taught the required material. The response from the Japanese teachers was that such a thought would be nearly unthinkable, and that "good teachers" would not think that. Doing *kyozaikenkyu* is one way for teachers to prepare themselves and a lesson that gives students the best chance to succeed. A Japanese teacher would feel bad if their students were not learning as much or as well as the students in other teacher's classes in the same school.

Adaptation to Students

Earlier we described a puzzle, that Japanese teachers still spend time crafting or developing their own lesson even with easy access to excellent "ready-made" resources. One answer to this puzzle is the idea that any published materials are written for a general class, not for a specific class, and that the lesson can be much better if a teacher prepares for their particular students. The following response is from a Japanese mathematics educator that was interviewed as part of our larger study about the importance of adaption:

> *Kyozaikenkyu* should be done for today's students, their students. When they do their *kyozaikenku* they should have their real students in mind. On the other hand, the textbook is generalized, not about specific students. They need to adapt to their students, if they don't adapt the textbook, it may not be good enough for their students. The information in the teacher's manual may be an average lesson, but it can be better if they do their own *kyozaikenkyu*, but some give up and just use textbook, and it will be an average just-OK lesson.

Japanese teachers seem to adapt lessons to improve their students' learning experience.

Mastering the Teaching Craft

This value is closely tied to the idea of *seishin* as well as *jutsu* and *waza*. Teaching mathematics is viewed as the teacher's craft, or *jutsu*, and to uncover the *waza* to that craft requires consistent and focused effort. It is viewed as noteworthy to work hard for a long time to be good at anything worthwhile (*seishin*). This may be especially true if that craft or activity is your profession. Through *kyozaikenkyu*, teachers are constantly learning about their craft by developing deeper understanding of mathematics, by understanding the ways students think and solve problems, by learning other ideas or strategies from their colleagues, by modeling lessons like those of expert teachers, and by understanding which small choices (like the phrasing of a question) can make a big difference in what and how students think and learn.

Detailed Lesson-Plan Writing: Behavior

On one occasion, as part of our larger study, two US teachers were preparing a set of special lessons on understanding matrix multiplication (what it means in a particular context and why we multiply matrices with what otherwise seems as an arbitrary procedure). The Japanese colleagues asked the US teachers to send them a lesson plan a few weeks before the Japanese cohort came to the US so they could be better prepared to learn from the lesson observation. The US teachers were at a loss about what to send. They did not know what the Japanese colleagues expected, and it is also not a common practice to write out lesson plans for colleagues for most US teachers. The US teachers sent a copy of the prepared handout, which the students would work from and fill out during the lessons. The handout included the stated tasks and some follow up questions. The Japanese were surprised that it was considered the lesson plan and did not find it very satisfying, asking again for a lesson plan to be sent to them.

Almost all Japanese mathematics teachers can develop a detailed lesson plan (a lesson plan like those shared with lesson study participants). It was very surprising for our Japanese colleagues that these two expert US teachers could not write a detailed lesson plan. What was also surprising was the fact that the lessons the Japanese observed were very good, so the US teachers were prepared and the lesson was well thought out. However, they still struggled to communicate their plan and thoughts to the Japanese observers, shocking our Japanese colleagues who equate careful preparation and writing a detailed lesson plan.

Unlike the vast majority of US teachers, Japanese teachers have ample opportunities to consume and develop detailed lesson plans. Experience in writing detailed lesson plans begins as pre-service teachers. For in-service teachers, lesson plans are a key tool in every step of the lesson study process and a fundamental way of sharing instructional knowledge with peers.

Detailed Lesson-Plan Writing: Values

Since detailed lesson plan writing is closely related to the practice of *kyozaikenkyu*, the values of *kyozaikenkyu* apply here as well, particularly the value of mastering one's craft. The best lesson plans tend to come from the most capable and experienced teachers. We share two more values that seem most pertinent to writing detailed lesson plans. One is based on a responsibility, and one on openness.

Responsibility for Community Improvement

Among Japanese teachers there is a culture of helping each other succeed. In schools the teachers have common offices so it is easy for teachers to freely help each other as they prepare their lessons and do *kyozaikenkyu*. Teachers benefit from teachers at other schools who have published lesson study lesson plans, written books for teachers, or published in teacher magazines. They also benefit from in-person interactions at lesson study conferences and teacher math circles. Teachers get help from other teachers, especially when they are young. Recall that one of the cultural values of Japan mentioned earlier is *giri*, the feeling of obligation one should feel when someone does something nice to you. Teachers naturally feel that they should give back to the community that has helped them so much. Engaging in lesson study and writing and sharing detailed lesson plans is a way to contribute back to the community.

Openness of Teaching Practice

Related to the previous value of contributing to the improvement of the community is the value of openness of one's practice. Without a willingness to share ideas, to observe others teach, to have others observe you, and to collaborate with colleagues then little improvement could be made, both individually as well as a community. This is most prominent in lesson study, but exhibits itself in other ways as well. Sharing detailed lesson plans is one way to open up your thinking as a professional to others in the profession.

Confirmation Study Results

The results of our confirmation study are displayed below in Table We have tabulated the responses for each of the four ratings: strongly agree, agree, disagree,

Table Survey results from 84 elementary and junior high Japanese teachers

Eight identified values	N	Strongly agree	Agree	Disagree	Strongly disagree	Mean
Logical thinking as a life skill	84	37	43	4	0	1.6
Deeper understanding of mathematics	83	11	50	21	1	2.1
Being true to the mathematics discipline	84	35	44	4	1	1.7
Responsibility for student learning	82	54	27	1	0	1.4
Adaptation to students	81	46	35	0	0	1.4
Mastering the teaching craft	82	52	28	2	0	1.4
Responsibility for community Improvement	82	42	39	1	0	1.5
Openness of teaching practice	81	46	32	3	0	1.5

and strongly disagree. We have also displayed the mean of the responses, with SA = 1, A = 2, D = 3, SD = 4. There were very few negative responses and all of the averages were less than 2 except for one value: Deep Understanding of the Mathematics. Many of the teachers that marked disagree or strongly disagree for the value "Deep Understanding of Mathematics" commented that procedural skills and doing well on exams was also important, not just a deep understanding. Almost half of the teachers (7 of 15) that explained why they disagreed with the statement stated or implied in their comment that they personally agreed, but either both were important or "unfortunately" teachers must focus on tests because of parents and the reality of entrance exams.

Discussion and Conclusion

We have pointed out eight potential values of the Japanese mathematics teacher community. We do not claim that these are all of the important values of the Japanese teaching community, but they do seem fundamental in their influence on core community practices of Japanese mathematics teachers. The contrast on the confirmation study between deeper mathematical understanding and proficient exam scores actually helped to point to another possible community value: high performance on entrance exams. However, this will need further research, since this was outside the focus of the larger, long-term study. The comments also were inconsistent, with some indicating that high exam scores was not a core value of Japanese teachers, but of parents. However, some teachers said that they feel obligated to emphasize skills in their instruction at a level that sacrifices deep understanding to support parents and students quest for high scores.

We have emphasized looking at values at different levels. Although values can vary between individuals, we think it is valuable to think of values of communities and values of cultures. The values communities hold can strongly influence the values of individuals that are enculturated into that community. This phenomenon fits our experience studying Japanese teachers.

Is knowing specific community values of Japanese teachers worthwhile or meaningful? Well, they may be when trying to take effective practices from one culture/community to another. Implementing a practice in a new community where values are different will not likely have the same results (for a discussion about this issue in the context of lesson study see Fernandez and Yoshida 2004). Implementing lesson study, for example, in a country where values such as deeper understanding of mathematics, responsibility for community improvement, mastering the teaching craft, and openness of teaching practice does not exist will probably produce a different outcome in the nature and quality of the teacher interactions as well as the lesson. It may be reasonable to think of the values of a community as part of the treatment of the effect of lesson study (on teacher learning and instructional quality, for example), not just the practice or behavior. If this is in fact the case, then implementing lesson study outside of Japan would probably need to consider how to change cultural norms among teachers, not just practice. This is a far-reaching possibility to which further research attention needs to be directed.

References

Baba, T., Iwasaki, I., Ueda, A., & Date, F. (2012). Values in Japanese mathematics education: Their historical development. *ZDM Mathematics Education, 44*(1), 21–32.

Bishop, A. (1988). *Mathematical enculturation: A cultural perspective on mathematics education*. Dordrecht, The Netherlands: Kluwer Academic Publishers.

Bishop, A. J., Seah, W. T., & Chin, C. (2003). Values in mathematics teaching—The hidden persuaders? In A. J. Bishop, M. A. Clements, C. Keitel, J. Kilpatrick, & F. K. S. Leung (Eds.), *Second international handbook of mathematics education* (pp. 717–765). Dordrecht: Kluwer Academic Publishers.

Boyatzis, R. E. (1998). *Transforming qualitative information*. Thousand Oaks, CA: Sage.

Cai, J., & Garber, T. (2012). Teaching values and valued teaching in the mathematics classroom: Toward a research agenda. *ZDM Mathematics Education, 44*(1), 91–97.

Corey, D. L., Peterson, B. E., Lewis, B. M., & Bukarau, J. (2010). Are there any places that students use their heads? Principles of high-quality Japanese mathematics instruction. *Journal for Research in Mathematics Education, 41*, 438–478.

Fernandez, C., & Yoshida, M. (2004). *Lesson study: a Japanese approach to improving mathematics teaching and learning*. Mohwah, NJ: Lawrence Erlbaum Associates.
Katagiri, S. (2004). *Suugakutekina kangaekatano gutaikato shido (Mathematical thinking and how to teach it)*. Tokyo: Meijitosho. (in Japanese).
Kluckhohn, C. (1962). Values and value-orientations in the theory of action: An exploration in definition and classification. In T. Parsons & E. A. Shils (Eds.), *Toward a general theory of action* (pp. 388–433). New York: Harper and Row.
Law, H. Y., Wong, N. Y., & Lee, L. N. Y. (2012). A study into espoused values in Hong Kong mathematics classrooms. *ZDM Mathematics Education, 44*(1), 45–57.
Mellville, M. (2017). *Kyozaikenkyu: An in-depth look into Japanese educator's daily planning practices*. Unpublished Masters Thesis. Brigham Young University, Utah.
Rokeach, M. (1973). *The nature of human values*. New York: The Free Press.
Seah, W. T., & Wong, N. Y. (2012). What students value in effective mathematics learning: A 'Third Wave Project' research study. *ZDM Mathematics Education, 44*(1), 33–43.
Skemp, R. R. (1976). Relational understanding and instrumental understanding. *Mathematics Teacher, 77,* 20–26.
Wiersbicka, A. (1991). Japanese key words and core cultural values. *Language in Society, 20*(3), 333–385.

11

Valuing in Mathematics Learning Amongst Ghanaian Students: What Does it Look Like Across Grade Levels?

Ernest Kofi Davis, Monica E. Carr and Ernest Ampadu

Abstract In this chapter, the authors explored the effects of grade levels on what students find important in their mathematics learning in Ghana. A survey involving 1,256 primary, junior high and senior high school students was conducted in the Cape Coast Metropolis of Ghana, using the WIFI questionnaire. It revealed that the Ghanaian students valued attributes such as achievement, relevance, fluency, authority, the use of ICT, versatility and Strategies in their learning of mathematics. The one-way Multivariate Analysis of Variance (MANOVA) was used to investigate whether significant differences exist in what students valued in mathematics across grade levels. The results revealed a significant effect of grade level on students' valuing in mathematics. Implications for research and curriculum delivery are provided.

Keywords Values · Mathematics · Learning · Grade levels · Ghana

Introduction

The past fifteen years have seen growing interest and research activities around the world into the role of values in mathematics education. Initial research conducted through the lens of values and valuing involved small scale studies of teacher valuing in mathematics education (Chin and Lin 2000). The inception of the Third Wave Project in 2008 facilitated the large scale study of valuing in mathematics on what students value in their mathematics learning (Seah and Wong 2012). This created

E. K. Davis (✉)
School of Educational Development and Outreach, College of Education Studies,
University of Cape Coast, Cape Coast, Ghana
e-mail: ekdavis@ucc.edu.gh

M. E. Carr
The University of Melbourne, Melbourne, VIC, Australia

E. Ampadu
University of Ghana, Accra, Ghana

an opportunity to explore what students from poor countries that persistently perform poorly in international comparative educational assessment in mathematics and science (TIMSS) such as Ghana value in their learning of mathematics at school.

Ghana has participated in the last four TIMSS exercises, that is, in 2003, 2007, 2011, and 2015. The TIMSS (2011) reports have revealed that eighth grade students from Ghana performed at the lowest level when compared to grade matched students from the other participating countries (Mullis et al. 2012). Little research has been conducted in under-performing countries in international mathematics assessments such as TIMSS. Although most efforts to date have focused on high-performing countries, conducting research across the spectrum of performance is of equal importance. It is particularly important to note that values in mathematics education are believed to be related to the socio-cultural context of mathematics education (Bishop 2008). With this in mind, and in the spirit of inclusiveness, Ghana was invited to take part in the international 'What I Find Important (in my mathematics learning)' [WIFI] in 2015.

A survey of the 1,256 Ghanaian primary, junior high school (JHS) and senior high school (SHS) students who participated in the WIFI study revealed that they valued *achievement, relevance, fluency, authority, ICT, versatility, learning environment, strategies, feedback, communication, fun, connections, engagement, applications, accuracy* (Seah et al. 2017b). However, the data were not designed to be analysed to compare the valuing among students in the higher grade levels.

Valuing across grade levels is positioned as being important in this chapter because it sheds light on how valuing amongst students operating at various cognitive development levels compare. It also provides an avenue to relate the socio-cultural context of school at these levels to the valuing in mathematics education. In this chapter, our aim, therefore, is to contribute to the literature and practice on the effect of grade level on valuing in mathematics, by further exploring the Ghanaian data to understand whether grade level has an effect on what students find important in their mathematics learning. Exploring the effect of grade level is important to us because the stress or delight as students engage with more complex maths as they progress from one level to another in the school system is likely to bring out different values. In turn, these differences in students' enjoyment of mathematics may have some bearing on teaching techniques that are required to achieve optimal outcomes for all students.

We set the context for our study by providing a brief overview of the location of Ghana and mathematics education at the pre-tertiary level, a brief review of literature in the area of values/valuing in mathematics and mathematics education, and valuing across grade levels. Research methods and data collection procedures are described. Results on valuing across grade levels, discussion and conclusions are presented along with statements of implications for research and policy and practice.

School Mathematics in Ghana

Ghana is located on the west coast of Africa. It was a British colony named Gold Coast. The country covers an area of 238,534 km^2, with an estimated population in 2016 of 28.21 million (World Bank 2017). English language is the only national and official language of Ghana, although more than 50 indigenous languages are being spoken across the country. As with many sub-Saharan African countries, Ghana is a poor country with GNI per-capita of US$1380 (2016 estimate) (World Bank 2017).

Ghana presently practices a 14-year pre-tertiary system of education, comprising two years of pre-school, six years of primary and six years of secondary school education. Children are expected to start pre-school at age four, and primary school at age six. During the first half of secondary school JHS students studying grades 7, 8, and 9 are typically aged between 12 and 15 years. During the second half of secondary school SHS students studying grades 10, 11, and 12 are typically aged between 16 and 19 years. This pre-tertiary level education is presently free and compulsory, and as such the rate of student enrolment is very high. Differentiation between tracks such as general science, arts and technical starts at the SHS level.

The study of mathematics is compulsory in all years of the pre-tertiary level, and the mathematical contents covered in the school curriculum at this level is comparable to those covered in countries all over the world. All mathematics textbooks are written in English, although from pre-school to primary three, pupils are expected to learn mathematics in the local language. From primary four onwards, they are expected to learn mathematics in English.

All SHS students study core mathematics while those who are studying science-related courses take an additional mathematics course (elective mathematics). The use of information and communication technology in mathematics curriculum delivery in Ghana is limited (Mereku and Mereku 2014). Although the primary and JHS school mathematics syllabus recommend the use of calculators as a learning tool, students have very limited exposure to them. At these levels, calculators are prohibited in examinations. However, SHS school students have unlimited use of calculators, including in examinations and class exercises.

Students take internal examinations at the end of every term (approximately every 14–15 weeks), and external/national examinations organised by the West African Examination Council (WAEC) at the end of grades 9 (JHS three) and 12 (SHS three) respectively. WAEC is the regional examination body responsible for the assessment of the attained curriculum in the English speaking West African countries. Examinations at grade nine are meant for selection of students to the various programmes at the SHS school level. Examination at grade twelve is for selection of students to the various programmes at the university level. Stakes are usually very high in these examinations, and this often results in a situation where teachers have to teach to the test rather than employing approaches that enable students to acquire a conceptual understanding of concepts in attempts to finish the mathematics curriculum. The system is therefore often criticised as being examination dominated.

Primary mathematics teachers are generalist educated, in the sense that they teach all the subjects (including music and physical education) at the primary school level. However, JHS and SHS mathematics teachers are specialised. The minimum qualification for teaching at the primary and JHS level is currently Diploma in Basic Education, while the minimum qualification for teaching mathematics at the SHS level is either B. Ed (Mathematics) or BSc (Mathematics) with post graduate certificate in education. However, it is common to find teachers with qualifications other than these two teaching at the SHS level due to shortage of qualified teachers in some districts.

Values in Mathematics Education

Mathematical and Mathematics Educational Values

In this section, we will focus mainly on what Bishop (1996) called mathematical and mathematics educational values because of the key roles which the two categories of these values play in the quality of students' learning experience (Seah et al. 2017a) and for that matter in the quality of their learning outcomes. Although we are aware of other general educational values, that relate to norms of the educational institutions, such as honesty and punctuality (see for example Bishop 1988, 2008), these will not form part of the key discussion in this section. While brief highlights of these values may be described herein to illuminate the current research problem, greater detail may be found in past and present literature on these values (Bishop 1988, 2008; Dede 2006, 2015; Seah et al. 2017a).

Values appear to mean different things in different contexts (Dede 2006) to different people. They could represent a variable in an equation in one context such as in the study of school mathematics. Or it could mean the worth of something if used in an everyday conversation involving commerce. In this chapter, we are looking at values from Mathews' (2001) perspective, that is, as mediators of learning behaviours. This perspective helps us to position what students find important in their study of mathematics as something that mediates their learning behaviour in mathematics.

Values have been described as the main dependent variable in the study of culture, society, and personality and the main independent variable in the study of social attitudes and behaviour (Rokeach 1973). While values have been widely researched in many disciplines, as recently as 2003 the exploration of the role of values in mathematics education was described as sadly neglected (Bishop et al. 2003). Meanwhile, literature suggests that it constitutes an important variable in mathematics curriculum delivery in schools (Bishop 2008).

Values in mathematics education have been defined as attributes of importance and worth that are internalised by an individual that provides him or her with the will and determination to maintain any course of action chosen in the learning and teaching of mathematics (Seah and Andersson 2015). Values in mathematics education are believed to regulate the way in which a student's or teacher's cognitive skills and emotional dispositions are aligned to learning or teaching aspects of mathematics (Seah et al. 2017a; Seah and Andersson 2015).

Different researchers have categorized values in mathematics and mathematics education differently (see Bishop 1988, 2008; Dede 2011, for examples). Bishop (1988) conceptualised the categories of mathematical values and mathematics educational values and argued that mathematical values are the values espoused by Western Mathematics. Bishop (2008) posits they are "values which have been developed as the knowledge of mathematics has developed within 'Westernised' cultures" (p. 83). These values convey to students the subtle message of what is important in the process of mathematizing (Seah et al. 2017a). Hence, mathematics educational values are embedded in mathematics curriculum, textbooks and mathematics classroom practices (Bishop 2008). These values also convey the subtle message to students of "what it takes to learn mathematics well" (Seah et al. 2017a, p. 336).

Bishop (1988) categorizes three pairs of complementary mathematical values relating to ideological, sentimental and sociological values as *rationalism-objectivism*, *control-progress* and *mystery-openness* respectively. *Rationalism* relates to mathematical argument, reasoning, logical analysis and explanation. On the other hand, *objectivism* relates to objectifying, concretising, symbolising and applying mathematical ideas. Valuing *Control* relates to mastery of rules, facts, procedure and established criteria, while *Progress* relates to valuing of alternative methods, development of new ideas and questioning of existing ones. For the last complementary pair, the valuing of *Openness* relates to valuing of proofs and individual explanations, while the valuing of *Mystery* relates to valuing of the wonder, fascination and the mystique of mathematical ideas (Bishop 2008, p. 84). Thus, as Bishop (1988) argues, that these pairs of values are complementary suggests the need for mathematics education to promote the growth of both for effective mathematical training. These categorisations of values provide us with the theoretical perspective to investigate what students at the various grade levels find important about both mathematics and the process of learning it.

What students and teachers value in mathematics relate to the socio-cultural context of mathematics education (Bishop 2008). It may be possible to have students from two or more different contexts of schooling valuing different attributes in mathematics (Seah et al. 2017a). Therefore, investigating students valuing in mathematics across context has the potential to provide insights about how socio-cultural context of education at these levels might influence what they find important in "the practice of doing mathematics" and "what it takes to learn mathematics well" (Seah et al. 2017a, p. 336). In turn, this has the potential to unearth possible dominant pedagogical practices in the school system.

In this section we have provided a brief review of the literature on valuing in mathematics and mathematics education, highlighting the various categorisations of values in mathematics and mathematics education. We have also highlighted the socio-cultural nature of values in mathematics education and positioned research studies on effect of valuing across grade levels are being necessary. In the next section we would look at what is known in the literature about values across grade levels.

Values in Mathematics Education Across Grade Levels

There is a dearth of research on what students value in mathematics across grade levels. Some studies have looked at what primary and secondary school teachers value in mathematics/mathematics education and reported differences in values for primary and secondary school teachers (Bishop et al. 2005; FitzSimons et al. 2001; Dede 2015). For example, in a comparative study involving sixty primary and secondary teachers from Germany and Turkey, Dede (2015) found that school levels of the teacher within and between Germany and Turkey had a significant effect on their mathematics educational values. Since there are differences between teachers, it may well be that the same will apply to students.

Other values research studies in the area of psychology have also looked at what students value generally across grade levels. In one study of individual values focusing on learning the routine and academic procrastination of sixth and eighth grade students in Germany, Dietz et al. (2007) found amongst other results that "6-graders appreciate achievement values more than 8-graders" (p. 10).

Studies on values across grade levels help to ascertain the effect of context of mathematics education across these levels on valuing in mathematics education since Mathematics curriculum delivery across these levels in many countries are not the same. For example, who qualifies to teach mathematics, what the mathematical training background of teachers is, and whether students choose to do more difficult mathematics, could be very different across schools. We therefore posit that school context has the tendency to affect what students value in their mathematics learning.

The Research Context

The WIFI study is an international research project involving 20 countries including Ghana, across five continents namely Africa, Asia, America, Australia and Europe (Seah et al. 2017b) at the time of this study. The project employs a quantitative approach involving questionnaire survey to investigate what students find important in their mathematics learning. Data collection has been conducted by local teams of researchers from the participating countries. Detailed current literature on the WIFI project and the justification of its methodologies is well documented in several studies (e.g. Seah et al. 2017a). In this chapter we will focus our discussion on how the WIFI methodology was adopted in the investigation of valuing by Ghanaian students at various grade levels. The research questions "What do primary, JHS, and

SHS students' value in their mathematics learning?" and "How similar or different are mathematics students' values at primary, JHS, and SHS levels?"

Research Instruments

The validated questionnaire used for the WIFI Study was used to collect data in this study. The questionnaire is made up of four sections. The Likert-type scoring format was used for the first 64 items in Section A, in which students were asked to indicate how important mathematics pedagogical activities such as small-group discussions (item 3), connecting mathematics to real-life (item 12) and mathematics homework (item 57) were to them. A five-point scoring system was used, ranging from absolutely important (1 point) to absolutely unimportant (5 points). Section B consisted of continua dimensions, each related to two bipolar statements and respondents were asked to indicate along the continuum the extent to which their valuing leans towards one of the two statements. Section C consisted of four scenario-stimulated items, and Section D elicited the biographic data of the students. The English version was administered to Ghanaian students because English is the medium of instruction from grade four onward in Ghana. The questionnaire can be accessed at: https://melbourneuni.au1.qualtrics.com/jfe/form/SV_6YDuI41EnRFvozz.

Participants

The research participants in this representative sample were drawn from Government-funded public schools at the primary, JHS and SHS levels in the Cape Coast Metropolis of Ghana. A stratified random sampling procedure was used to select students from a mix of schools by achievement levels (Above Average, Average, Below Average) and by school context (i.e. rural and urban). The final set of 1,256 student participants consisted of 414 primary four, five and six, 426 JHS and 416 SHS students. The research participants attended 18 of the total possible 76 Government-funded public schools, whose enrolment totalled 42,257 students at the time this research was conducted.

Data Analysis

To answer the research questions: "What do primary, JHS, and SHS students' value in their mathematics learning?" and "How similar or different are mathematics students' values at primary, JHS, and SHS levels?"; a one-way Multivariate Analysis of Variance (MANOVA) was carried out. The MANOVA analysis afforded the opportunity to investigate the effect of grade level on valuing in mathematics, within acceptable error margin. A principal component analysis (PCA) with a varimax rotation and Kaiser normalization had been used to examine the questionnaire items, with significance level set at 0.05, while a cut-off criterion for component loadings of 0.45 was used in interpreting the solutions. Items that did not meet the criteria were eliminated. According to the cut-off criterion, 23 items were removed and 41 items were retained from the original 64. The analysis yielded 15 components with eigenvalues greater than one, which accounted for 52.73% of the total variance. Fifteen components of the students' set of values for mathematics learning were identified (Seah et al. 2017b). However, in MANOVA analysis, all components that had fewer than two item loadings were treated as weak factors and therefore excluded. This reduced the number of components from fifteen to seven: namely *C1-Achievement* (knowing the steps to solution, doing a lot of mathematics work etc.)*; C2-Relevance* (stories about mathematics, explaining where rules/formulae came from etc.)*; C3-Fluency* (practicing how to use maths formulae, explaining my solution to class)*; C4-Authority* (explaining by the teacher, learning maths with computer)*; C5-ICT* (using calculator to calculate, using calculator to check the answer)*; C6-Versatility* (being lucky at getting the correct answer, looking for different possible answers)*; C7-Strategies* (given a formula to use, shortcuts to solving mathematics problems). Grade levels formed the independent variable while the seven factors formed the dependent variable. Prior to the analysis, the data were screened to test for multivariate normality, homogeneity of covariance matrices (using Box's M test) and independence of observations. The details of the results obtained from the data analysis and the ensuing discussions, conclusions and implications are presented in the sections that follow.

Results

In this section, the MANOVA results of the 414 primary, 426 JHS and 416 SHS students who participated in the WIFI Ghana study are provided. The existence of statistically significant differences between primary, JHS and SHS students for each of the seven components *(C1-Achievement, C2-Relevence, C3-Fluency, C4-Authority, C5-ICT, C6-Versatility, C7-Strategies)* derived from the PCA, were investigated.

Pillai's Trace criterion was used to test whether there are significant group differences on a linear combination of the dependent variables. Since the multivariate effect for grade level is significant [Pillai's Trace value = 0.417, F(14, 2496) = 46.932, $p < 0.001$, partial eta square = 0.208, the power to detect the effect = 1.000], we interpreted the univariate between-subjects effects by adjusting for family-wise or experiment-wise error using a Bonferroni-type adjustment, and we derived the adjusted alpha level 0.007 (i.e. 0.05/7) (Coakes and Ong 2011). Using this alpha level, we have significant univariate main effects for each of the seven components:

- *Component 1 (C1): achievement* [F(2, 1253) = 137.74, $p < 0.001$, *partial eta square*(η^2) = 0.180, *observed power* = 1.000]

- *Component 2 (C2): relevance* [F(2, 1253) = 33.187, p < 0.001, *partial eta square* $(\eta^2) = 0.050$, *observed power* = 1.000]
- *Component 3 (C3): fluency* [F(2, 1253) = 27.608, p < 0.001, *partial eta square* $(\eta^2) = 0.042$, *observed power* = 1.000]
- *Component 4 (C4): authority* [F(2, 1253) = 10.266, p < 0.001, *partial eta square* $(\eta^2) = 0.016$, *observed power* = 0.934]
- *Component 5 (C5): ICT* [F(2, 1253) = 45.368, p < 0.001, *partial eta square* (η^2) = 0.068, *observed power* = 1.000]
- *Component 6 (C6): versatility* [F(2, 1253) = 7.809, p < 0.001, *partial eta square* $(\eta^2) = 0.012$, *observed power* = 0.826]
- *Component 7 (C7): strategies* [F(2, 1253) = 46.437, p < 0.001, *partial eta square* $(\eta^2) = 0.069$, *observed power* = 1.000].

Tukey's Honestly Significant Difference (HSD) Post Hoc multiple comparisons test was conducted to further determine which differences are the sources of the significant *F*-ratio obtained for the overall MANOVA, that is, between which of the three groups (primary, JHS and SHS students) and the seven dependent variables (C1, C2, C3, C4, C5, C6 and C8) are significant differences found. Tukey's HSD Post Hoc test uses the harmonic mean sample size for unequal group sizes. Significant grade level pairwise differences were obtained in what students valued between primary students and both JHS and SHS students. The analysis showed that there are statistically significant differences between primary and JHS students for six out of the seven components (C1, C2, C3, C4, C6 and C7) and primary and SHS students for four out of the seven components (C1, C2, C5 and C7). Statistically significant differences were also found between JHS and SHS students for four out of the seven components (C2, C3, C5 and C6). The estimated means for statistically significant components are indicated below:

- The primary students had the highest mean, followed by JHS students and SHS students (lowest) in Components 1, 5 and 7. In other words, the SHS students valued C1, C5 and C7 more than their peers at other school levels;
- The primary students had the highest mean, followed by SHS students and the JHS students (lowest mean) in Components 3, 4 and 6;
- The SHS students had the highest mean, followed by the primary and JHS students (lowest mean) in Component 2.

In summary, a one-way MANOVA revealed a significant multivariate main effect for grade level [Pillai's Trace value = 0.417, F(14, 2496) = 46.932, p < 0.001, partial eta square = 0.208, the power to detect the effect = 1.000]. Given the significance of the overall test, univariate main effects were examined. Significant univariate main effects were obtained for each of the seven dependent variables. Table presents the summary of grade level pairwise differences among the various grade levels and their significance level.

Table shows that there were significant grade level pairwise differences in C1, C2, C3, C4, C6 and C7 between primary and JHS students, C1, C2, C5 and C7 between primary and SHS students and C2, C3, C5 and C6 between SHS and JHS students.

Table: Grade level pairwise differences

Dependent variable	Mean diff Pri/JHS	P-value	Mean diff Pri/SHS	P-value	Mean diff JHS/SHS	P-value
Achievement-C1	0.5115	0.000	0.5898	0.000	0.0783	0.103
Relevance-C2	0.1982	0.000	−0.2036	0.000	−0.4017	0.000
Fluency-C3	0.5264	0.000	0.1211	0.238	−0.4053	0.000
Authority-C4	0.2639	0.000	0.1763	0.009	−0.0876	0.301
ICT-C5	0.0598	0.638	0.5755	0.000	0.5157	0.000
Versatility-C6	0.2090	0.001	0.0166	0.958	−0.1924	0.003
Strategy-C7	0.4503	0.000	0.5077	0.000	0.0575	0.576

Discussion

This study investigated the effect of grade level on what Ghanaian students find important in their study of mathematics. The results suggest that there were statistically significant differences between primary, JHS and SHS students for seven components namely, *achievement, relevance, fluency, authority, ICT, versatility, and strategies*.

The results showed a decrease in the estimated marginal mean scores for grade level from primary, JHS and SHS students respectively for the valuing of *achievement* (doing a lot of mathematics work, knowing the steps to solution, knowing which formula to use, understanding why a solution is incorrect or correct), *ICT* (using calculator to check the answer, using calculator to calculate) and *strategies* (short cuts to solving mathematics problem, given a formula to use). These attributes relate to the mastery of mathematics content hence one can argue that primary school students value having mastery of mathematics content less than JHS and SHS students, while JHS students also value mastery of mathematics content lesser as compared to SHS students.

Valuing "given formula to use" and "shortcuts to solving mathematics problem" (*Strategies*) and "knowing the steps of solutions", "knowing which formula to use" and "understanding why my solution is incorrect or correct" (*Achievement*), relates more to what Bishop (2008) describes as "... power of mathematical knowledge through mastery of rules, facts, procedures and established criteria" (p. 85), which relates to sentimental value of control in values in western mathematics. Our results, therefore, suggest that valuing of control increases as one moves up grade levels.

The finding regarding the valuing of the use of ICT in mathematics (that is, using calculators to check the answer and using a calculator to calculate) is not surprising. Having been exposed to the use of calculators and other ICT tools more than primary and JHS students, one would expect the SHS students to appreciate the affordances they offer in the teaching and learning of mathematics than the primary and the JHS students. All (primary, JHS and SHS) students find mathematics difficult and see the calculator as a tool with which they can carry out calculations without necessarily recalling number facts or algorism. However, the primary and the JHS students have very limited exposure to the calculator and other ICT tools in their lessons, even though the mathematics syllabus recommends them to do so. At the SHS level, it is possible that the students have gone over this restrictive pre-SHS level belief about the use of ICT tools such as calculators just for checking results after one had applied an algorithm. They would likely have realised the other benefits of calculators and ICT tools in supporting their mathematics learning, such as promoting in-depth understanding of concepts. This is because at the SHS level students are exposed to the use of calculator and other ICT tools. Many of the SHS students might have now realised that the calculator and the other ICT tools play a major role in their success in mathematics learning so far. They might have begun to place more importance on the use of ICT. We argue that the socio-cultural context of mathematics education in Ghana has shaped the valuing of ICT among the SHS students and the primary and JHS students. The restrictive use of ICT by pre-SHS students in their mathematics learning as compared to the SHS students might have shaped what they value in their mathematics.

There was also a decrease in the estimated marginal means—and thus, an increase in valuing—from primary to JHS/SHS levels in the student valuing of *fluency* (explaining my solution to class, practicing how to use maths formula), *authority* (learning maths with internet and explaining by the teacher) and *versatility* (looking for different possible answers, being lucky at getting the correct answer). Both the JHS and SHS students valuing of attributes such as practising how to use mathematics formula, explaining by the teacher and being lucky to get correct answers more than Primary school students supports our earlier observation about increase in valuing of mastery/control over mathematics content as one moves up grade levels.

There was an upward trend with estimated marginal mean score increasing with grade level for the valuing of *relevance*. In fact, SHS students valued *relevance* in their mathematics learning less than both primary and JHS students. They valued attributes such as stories about mathematics, explaining where rules/formulae came from, the mystery of mathematics, stories about recent developments in mathematics, using concrete materials to understand mathematics less than both the primary and JHS students. These attributes generally relate to the sociological values of openness and mystery in values in western mathematics (Bishop 1988). Explaining where rules/formulae came from and using concrete materials to understand mathematics relates to *openness*. While stories about mathematics, stories about recent developments in mathematics and mystery of mathematics relate to the *mystery* (Bishop 2008, p.85).

This shows that SHS students value sociological values of *openness* and *mystery* less than both primary and JHS students. The SHS students are more interested in mastery than understanding mathematics. This may be because at the SHS level the breadth and depth of mathematics contents that have to be covered are high and the pressure of high-stake national examinations may be a contributory factor.

Conclusions and Implications

This study has contributed to our understanding of how students' valuing develops across grade levels in Ghana. We investigated whether grade level had an effect on what primary, JHS and SHS students value in mathematics. A one-way MANOVA revealed a significant multivariate main effect for grade level [Pillai's Trace value = 0.417, $F(14, 2496) = 46.932$, $p < 0.001$, partial eta square = 0.208, the power to detect the effect = 1.000]. Given the significance of the overall test, univariate main effects were examined, and this also revealed significant univariate main effects for each of the seven dependent variables namely *achievement, relevance, fluency, authority, ICT, versatility* and *strategies*.

We, therefore, conclude that interesting differences in valuing in mathematics education across grade levels do exist for primary, JHS and SHS students. The mathematical value of control (Bishop 1988) seems to be embraced more by students in higher grade levels from JHS to SHS. The SHS school students valued *achievement, ICT* and *strategies* more than JHS and Primary school students, while the JHS students also valued these attributes more than their peers in Primary schools. Thus, we might suggest that values evolve within each person at least throughout the schooling years, if not over an even longer period of time.

The findings from this study appear to suggest that all the seven values are embraced by the students, except that some of them are more highly prioritised at different stages of schooling. For example, while control is valued at all levels, it is prioritised at the SHS level as compared to the Primary and JHS levels.

The findings from this study have implications for further research. In order to provide a better understanding of valuing across grade level, it might be important to ascertain through research, not only on how students' values reflect those of their teachers but also how values espoused by curriculum materials such as the syllabus and textbooks reflect those of the students. Future studies, such as the WhyFI study being led by a Hong Kong team, are using methodologies that investigate the reasons behind observed trends emerging from large scale surveys such as the one reported here. Although the sample for the study was large enough to support generalisation of our findings, it is important to state that the study was conducted in only one out of ten regions in Ghana. The findings might therefore not reflect the entire situation in Ghana. Hence future studies might also include samples from the other regions in the country.

References

Bishop, A. J. (1988). *Mathematical enculturation a cultural perspective on mathematics education*. Dordrecht, The Netherlands: Kluwer.

Bishop, A. J. (1996, June 3–7). How should mathematics teaching in modern societies relate to cultural values—some preliminary questions? In *Paper presented at the Seventh Southeast Asian Conference on Mathematics Education, Hanoi, Vietnam.*

Bishop, A. J. (2008). Teachers' mathematical values for developing mathematical thinking in classrooms: Theory, research and policy. *The Mathematics Educator, 11*(1/2), 79–88.

Bishop, A. J., Clarke, B., Corrigan, D., & Gunstone, D. (2005). Teachers' preferences and practices regarding values in teaching mathematics and science. In P. Clarkson, A. Downton, D. Gronn, M. Horne, A. McDonough, R. Pierce, & A. Roche (Eds.). *Building connections: Theory, research and practice, proceedings of the 28th annual conference of Mathematics Education Research Group of Australasia,* (pp. 153–160). Melbourne, MERGA.

Bishop, A. J., Seah, W. T., & Chin, C. (2003). Values in mathematics teaching—the hidden persuaders? In A. Bishop, M. A. Clements, C. Kietel, J. Kilpatrick, & F. Leung (Eds.), *Second international handbook of mathematics education* (pp. 717–765). Dordrecth: Kluwer Accademic Publishers.

Chin, C., & Lin, F. L. (2000). A case study of a mathematics teacher's pedagogical values: Use of a methodological framework of interpretation and reflection. http://140.122.140.1/~cchin/file01.pdf.

Coakes, S. J., & Ong, C. (2011). *SPSS verion18 for windows: Analysis without anguish.* Milton, Qld, Australia: Wiley.

Dede, Y. (2006). Values in Turkish middle school mathematics textbooks. *Quality & Quantity, 40,* 331–359.

Dede, Y. (2011). Mathematics education values questionnaire for Turkish preservice mathematics teachers: Design, validation and results. *International Journal of Science and Mathematics Education, 9*(3), 603–626.

Dede, Y. (2015). Comparing primary and secondary mathematics teachers' perferences regarding values about mathematics teaching in Turkey and Germany. *International Journal of Science and Mathematics Education, 13,* 227–255.

Dietz, F., Hofer, M., & Fries, S. (2007). Individual values, learning routines, and academic procastination. *British Journal of Educational Psychology, 77,* 893–906.

FitzSimons, G., Seah, W. T., Bishop, A. J., & Clarkson, P. (2001). Beyond numeracy: Values in mathematics classroom. In J. Bobis, B. Berry, & M. Mitchelmore (Eds.). *Numeracy and beyond. Proceedings of the 23rd Annual Conference of Mathematics Education Research Group of Australasia,* (pp. 202–209). Fremantle: MERGA.

Mathews, B. (2001). The relationship between values and learning. *International Education Journal, 2*(4), 223–232.

Mereku, D. K., & Mereku, C. W. K. (2014). *Congruence between the intended, implemented and attained ICT curricula in sub-Saharan Africa.* Online: www.tandfonline.com/dpi/abs/10.1080/14926156.2014.992555.

Mullis, I. V. S., Martin, M. O., Foy, P., & Arora, A. (2012). *TIMSS 2011 International Results in Mathematics.* Chestnut Hill, MA. https://timssandpirls.bc.edu.

Rokeach, M. (1973). *The Nature of human values.* New York: Free Press.

Seah, W. T., & Andersson, A. (2015). Valuing diversity in mathematics pedagogy through the volitional nature and alignment of values. In A. Bishop, H. Tan, & T. Barkatsas (Eds.), *Diversity in mathematics education: Towards inclusive practices* (pp. 167–183). Switzerland: Springer.

Seah, W. T., Baba, B., & Zhang, Q. (2017a). The WIFI study students valuing in Hong Kong and Japan. In J. W. Son, T. Watanabe, & J. Lo. (Eds). *What matters? Research Trends in International Comparative Studies in Mathematics Education, Research in Mathematics Education* (pp. 333–354). New York, Springer.

Seah, W. T., Davis, E. K., & Carr, M. E. (2017b). School mathematics education through the eyes of students in Ghana: Extrinsic and intrinsic valuing. In *10th Congress of European Research in Mathematics Education.* https://minerva-access.unimelb.edu.au.

Seah, W. T., & Wong, N. Y. (2012). What students value in effective mathematics learning: a 'Third Wave Project' research study. *ZDM: The International Journal on Mathematics Education, 44,* 33–43.

World Bank. (2017). *Ghana data.* https://data.worldbank.org.

12

Democratic Actions in School Mathematics and the Dilemma of Conflicting Values

Annica Andersson and Lisa Österling

Abstract This chapter reports and problematizes relationships between the expected democratic actions as part of the politically expected democratically inclusion of students' wishes and concerns; and students' valuing of mathematical activities in mathematics classrooms, departing from the Swedish results from a large-scale quantitative cross-cultural survey. We asked what are the conflicts between most valued activities by Swedish students and the valuing of democratic actions. The quantitative study showed that students value "knowing the times tables" and "teachers' explanations" and "correctness" over explorative, communicational and collaborative activities. We discuss the cultural and historical reasons behind these results and argue that we must understand the valuing of times tables or teachers' explanations as an expression of enculturated and therefore culturally valued actions in mathematics classrooms, where this enculturation takes place not only in school, but in conversations with parents, grandparents, in media and in children's books. We also argue that the conflict between the political expectations of democratic participation and actions, and the invitation to students to influence teaching on the one hand, and on the other hand students use of this influence through valuing teacher explaining, mastering times tables and understanding why the answer is incorrect, rather conserve a mathematics teaching organised around values as objectism and control than through openness and rationalism.

Keywords Democracy · Enculturation · Mathematical activities · Mathematics education · Students' valuing

A. Andersson (✉)
University of South-Eastern Norway (USN), Post Office Box 4, 3199 Borre, Norway
e-mail: annica.andersson@usn.no

L. Österling
Stockholm University, Stockholm, Sweden
e-mail: lisa.osterling@mnd.su.se

Introduction

This chapter reports and problematizes the relationships between the expected democratic actions and values in mathematics classrooms, departing from the Swedish results from the large-scale quantitative cross-cultural survey, 'What I Find Important (in mathematics learning)' [WIFI]. It surveys students' valuing of different mathematics learning activities and their importance for students' learning across cultures. Initially, two main reasons made us use the WIFI questionnaire. First, we wanted to learn about what Swedish students value when participating in mathematical activities, with a possible further goal to understand cultural similarities and differences from the comparisons of results from the participating 19 countries. This will be briefly discussed and problematised as the survey results are the background for this study. The second reason was to further understand students' valuing, as part of the politically expected democratically inclusion of students' wishes and concerns in the planning of mathematics classroom activities (Ministry of Education 2011). The latter reason is the foregrounded focus of this chapter.

In this chapter, we first give a theoretical and contextual research background. We then present the survey and discuss its validity before accounting for the results of the 748 eleven- and fifteen-year-old students' survey responses. Thereafter, based on these results, we address the challenges teachers may experience when working towards a politically expected democratic inclusion in mathematics education. In doing so, we take our departure with the assumption that teaching, as well as students' valuing of mathematical classroom activities is, of course, not only influenced by the recent curriculum. It is also influenced by ideologies and epistemologies of the learning of mathematics, as well as contextual and societal cultures and traditions.

Theoretical Framework

Mathematical Values and Democracy

The idea that *Mathematics* is a neutral and objective area of knowledge is challenged in research (c.f. Biesta 2009; Skovsmose 2005). For example, the historical but also recent changes in Japanese school mathematics illustrate how different aspects of mathematics has been both valued by, and depending on, societal and political changes (Baba et al. 2012). The work of Alan Bishop provided an important contribution to the understanding of values in mathematics. He defined mathematics learning as an enculturation process where students are expected to enter and become part of mathematics communities and thereby enculturate values of mathematics and mathematics education. In this process, teachers and students "do not have equal roles to play, nor do they meet on equal terms" (Bishop 1988). A legitimate use of teachers' power can, according to Bishop (2002), be the specific fostering of mathematical values in order to enculturate students into the culture of a mathematical community. This process might involve a shift of students' values; from the cultural values students bring to school, to the appreciated mathematical values necessary for the entrance to the mathematical community.

We here emphasise that we resonate with Hofstede et al. (2010) definition of values as shared by members of a community. Such values will not only be shared, but also part of determining who is a community member or who is not, since social conventions as well as laws are based on shared community values. Those values might or might not be recognized by the members of such communities, but value-differences from other communities may be easily recognized. Thereby, in this chapter we understand values as guiding our valuing of what is important to students and teachers (and researchers) or not, what is good or bad, what is beautiful or ugly, and so on. In order to choose to engage in purposeful learning, students and teachers in mathematics classrooms need to share some basic values of mathematics and mathematics education. In addition, we add, that to make students' democratic actions possible, spaces for conversations, negotiations and expressing views on and with mathematics and mathematical knowledge need to be available and negotiable in mathematics classrooms.

The basic general values that give political educational directions for teachers in Swedish schools are based on societal values as "fundamental democratic values" and "human rights" (Ministry of Education 2011, §4, 5). These notions grant students a large amount of influence on their education and also a responsibility, as being part of civil democratic obligations. Here are two examples from the national steering documents we refer to:

> Teachers shall:
>
> Be responsible for ensuring that all pupils can exercise real influence over working methods, forms and contents of education, and ensure that this influence grows with increasing age and maturity.
>
> /…/
>
> Prepare pupils for participating and taking responsibility, and applying the rights and obligations that characterise a democratic society (Skolverket 2011, p. 17, Skolverket's translation).

These paragraphs shall be taken into account in all subject areas, and are therefore not explicitly repeated in the mathematics part of the curriculum. Democratic participation is not to be understood as democratic education per se nor creating democratic citizens or individuals through teaching about democracy. Actually, the word "democracy" is not explicitly defined in the Swedish curriculum. Instead, it is based on values as explained in the very first paragraph:

> The Education Act (2010:800) stipulates that education in the school system aims at *pupils acquiring and developing knowledge and values*. […] *Education should impart and establish respect for human rights and the fundamental democratic values on which Swedish society is based*. Each and everyone working in the school should also encourage *respect for the intrinsic value of each person and the environment* we all share (Skolverket 2011, p. 9, our italics).

In an OECD report from 2006, we read that teachers should be "providing equal opportunity for all children within a universal system in which values of citizenship are inculcated, and a democratic and multicultural mixing of children is practiced" (p. 118) and specifically for the Nordic tradition focusing on democracy: "Centre goals are to support child development and learning and provide experience of democratic values" (p. 143). What these "democratic values" explicitly means is again not clear. However, researchers in educational values in early childhood education, points out that the values in Nordic countries may differ amongst each other. For example, Alvestad and Samuelsson (1999) showed that when the Norwegian curriculum was built upon a Christian orientation, the Swedish was built upon a more democratic perspective. A textual analysis of Nordic preschool curricula showed that

> In the Danish, Icelandic, Norwegian and Swedish curriculum frameworks, democracy is explicitly defined as one of the fundamental pillars that the guidelines are based on, and thus, the term is used frequently throughout the documents. The Finnish curriculum guidelines are unique in that they do not use the term democracy. However, basic notions of democracy, such as children's participation and influence, are stressed in all the documents and form a foundation for pedagogical practice (Einarsdottir et al. 2015, p. 102).

As an alternative to "democratic participation" we rather bring the idea of democratic actions (Biesta 2007) forward. Biesta argues that individuals, through actions as an alternative to participating in democratic education as a school subject, "become subjects in the routines of everyday [school] life". Hence students' experience of democracy is "lived", "becomes real" and thus "make [democratic] action possible" (p. 761).

In other words, our understanding of the political intentions formulated in the Swedish curriculum of "democratic participation", "inclusion", or as we prefer, "democratic action" (Biesta 2007) is that students shall experience opportunities to firstly, actively participate in planning, working and assessing activities in the mathematics classroom, and secondly, to participate in mathematical activities that to some extent are exploratory and open for collaborations, conversations and contributions of all participants on equitable premises.

Students' Democratic Participation in Mathematics

In a global context, the United Nations (2016) recognises democracy as one of its core values and ideologies. The quotes above from the Swedish steering document demonstrate that Swedish schools are not value-neutral. Swedish education aims to reinforce rights and obligations that characterise a democratic society. Thus, democracy here is connected to values as the means and purpose of democracy in line with the United Nations Declarations (2016) as to realize human rights.

These perspectives touch on many areas in mathematics education, from the shared responsibility for learning, over intertwinedness of human relations in the participation in mathematics work, and finally the power to decide what counts as mathematical knowledge. An extended literature review of research about democracy and mathematics education highlights the tensions within this relationship and concludes that "indeed there are connections between mathematics education and democracy,

[...] However, these connections are not always positive. Mathematics education can promote democratic competences and values, but it can also inhibit them, and create social inequalities" (Aguilar and Zavaleta 2012, p. 10)

From a student's perspective, Zizzi's, the democratic action as shared responsibility seems to be a new and strange experience in school mathematics.

> This was really meaningful and it was good to take personal responsibility for planning and for our own labour. But this is new; we have to practice this way of working. (Andersson and Valero 2015, p. 212)

Zizzi, 15, shared this comment on the classroom blog after working in a group in a mathematical project that was constructed with the aim of allocating students' responsibilities for planning and assessing the project while pursuing their mathematical knowledge. She acknowledged the mathematical goals of the project, and the responsibilities that came with this way of working. However, she pointed out that this was a new way for her to work in a mathematics classroom. Her previous experiences consisted of individual textbook work, and this comment alerted the teacher to realise that democratic or collaborative ways of working in mathematics needed to be learnt, if a shared responsibility between teachers and students should be attained in the mathematics classroom.

The fact that basic general democratic values give directions for teachers impose that Swedish schools grant students a large amount of influence on their education. However, several research examples demonstrate how challenging it may be to combine with mathematics learning. As part of a modelling research project in upper secondary schools, Lingefjärd and Meier (2010) analysed a classroom-vignette where a group of students wants to further question and discuss their developed advanced formula with their teacher, who responds: "Well, if it is your formula, then go ahead and explain it!" (p. 103). When the researchers ask further about this interaction, the teacher refers to the mathematics curriculum, which with this teacher's words stated that students should 'learn to work on their own'. However, what this episode shows are that despite the teacher's good intentions of granting students influence, the lack of framing resulted in these students neither learned mathematics nor learned to work independently.

In addition, Johansson (2006) described how, in the name of democratic participation, the responsibility for learning is often passed on from teachers to students and even to the textbooks. Especially students with Swedish as their second language are found to be disadvantaged in this type of teaching. Thus, instead of improving inclusion in mathematics education, this way of making students 'responsible' for their learning widens the gap between students who are familiar with the expectations and discourses within the mathematics educational culture, and students who do not have access to the discourses (Hansson 2010). Teachers, in the name of allocating influence and responsibility, instead engage in a form of individualized or student-centred learning, a practice where teachers are seen too often abandon students who need guidance and instead expect students to work on their own with textbooks tasks and problems (Johansson 2006; School Inspectorate 2009, 2010). In contrast, Norén (2015) demonstrates how teachers allowed second language learners taking control of their learning by using their previous experiences, and thus allowed them to become more engaged in mathematical activities. Here, the inclusion of learners was facilitated by adaptation to students' experiences.

To invite students to engage in a teaching based on democratic actions often includes collaborations not only with the teacher, but also with peers. Thus, human relations in group work becomes intertwined with the learning of mathematics. Kurth et al. (2002) reminds us that students might be unsuccessful due to their obligations both to the working group and to the learning of mathematics. These two aspects also occur when Wood (2013) describes grade four students' different positionings in mathematics group work (as experts, novices or facilitators) demonstrates the complexity of collaborative work. Esmonde (2009) showed a range of different work practices were individual students adopted these positions in three secondary classrooms over a year. The mathematical interactions were mainly dominated by the "experts" whereas "interactions were more equitable particularly when a student was positioned as a facilitator" (p. 247). As we interpret Esmonde's findings, more democratic actions might take place when a facilitator is positioned within mathematical working groups. Andersson and Wagner (2017) alert us through a SFL analysis how mathematical conversations underpin both "love" and "bullying" in students' interactions. In addition, DeJarnette and González (2015) show how mathematical reasoning and students' positionings in groups are intertwined, hence a democratically inclusion may depend on individual students' mathematical knowledge. In other words, some students may be more active in mathematical reasoning than others, who might be quieter. We add that democratic actions also call on aspects as listening and talking space. Thus, equitable premises are required in mathematical collaborative activities when aiming for democratic inclusion.

On the other hand, students' possibilities for achieving influence and acting in more democratic ways may be facilitated if students understand and agree on the stated mathematical and task objectives (Andersson 2011) or align with the prevailing values in the mathematical classroom (Seah and Andersson 2015; Swan 2014). The reasons for this might be explained by Wagner and Herbel-Eisenmann (2009, 2014). They make the distinction between students' personal authority, and the disciplinary authority of mathematics. The authority of mathematics might not lend itself to negotiation. Stemhagen (2016) argues, through using Dewey's ideas, where democratic participation and actions consists of mathematics linked to children's lived inquiries, that "deep and unresolved tensions in the philosophy of mathematics and the philosophy of education have made it difficult for promising [democratic] pedagogies to be enacted" (p. 95).

Nevertheless, democratic participation and actions, inclusion and collaborative work are politically expected virtues in mathematics education today, at least in Sweden, and hence require our research attention. From this overview, we conclude that an active participation of students, in the name of participation, positioning, authority, agency, activity or collaboration, is related to a democratic classroom. We also see how such classrooms might be specifically challenging for mathematics teachers to establish, or, as Bishop (1988) would put it, to enculturate democratic values. We remind ourselves about 15-year old Zizzi's comment, that organising mathematical activities democratically might need to be "practiced", or at the least negotiated between teachers and students.

Purpose and Research Questions

The purpose of this chapter is first, to understand what values students perceive as important in the Swedish mathematics classroom and second, to understand how students' valuing are in line with or opposed to values of democratic participation and actions. We relate to two guiding questions:

1. How important are different mathematics classroom activities, specifically those related to democratic actions, for Swedish students in their mathematics learning?
2. What are the conflicts between most valued activities by Swedish students and the valuing of democratic actions?

In other words, we explored whether if there is harmony, tension or a conflict between students' valuing of mathematical activities for learning mathematics and the politically desired democratic values in Swedish mathematics classrooms.

Methodology

The Survey Instrument

The WIFI study[1] was originally developed in English in an Australian-Asian context with the intention to learn more about what students' value when learning mathematics at school. This was obtained through a quantitative cross-cultural survey that investigated students' values through grading the importance of mathematical activities in more than twenty world-wide countries. However, children responding to the questionnaire cannot be expected to relate directly to a value; for example, it is difficult for students to understand or to answer the question "How important is rationalism when learning mathematics?" The participating students were instead asked to value 64 items describing various mathematical learning activities, by marking their importance for learning mathematics on a scale, from absolutely unimportant to absolutely important. In the WIFI-study, the intentions were to focus and compare as many activities as possible, to analyse these activities connectedness to certain mathematical values, and then compare the results between countries or cultures.

As we were conducting research in a Swedish language and context, we needed to address linguistic and cultural challenges at different stages of the project to be able to make cross cultural comparisons (Andersson and Österling 2014). A team of three researchers were engaged in a series of three adjugation meetings for such

[1]The survey is available at https://melbourneuni.au1.qualtrics.com/jfe/form/SV_7PaiTvst3sSflrL.

adaptation process, where the first meeting engaged the translated and back-translated questionnaire, the second the results from the piloted version, and the final third meeting developed the final version.

When we translated and adopted the questionnaire we followed the Survey Research Centre's (2010) "Guidelines for best practice in Cross-Cultural surveys". These guidelines state that translation and back translation processes are not sufficient, hence a cultural adaptation was necessary. Thus, direct translations are not always sufficient. Instead, adaptations to guide comprehension, improve conceptual coverage and a cultural sensitivity of what is appropriate are recommended. Therefore, we engaged a pilot-test, where 28 Swedish students of the targeted age-group were selected for interviews on their interpretation of difficult items. As an example, the English source questionnaire asked students to value the importance of "Mathematics debates". In the 1st translation, this was translated to "Debatter med matematik", and the back translation was close enough, "debating maths". However, when trying out the questionnaire in the pilot test, one out of three students did not understand the question. And when discussing "Mathematics debates" in the 2nd adjudication, not even the participating adjudicators were sure about how such a debate is enacted in the classroom. "Mathematics debates" are in the WiFi Research Guidelines (not published) classified as an indicator of valuing openness and exploration. Mathematics debates is not a common name of such activities in Swedish classrooms, so out of what it is supposed to indicate, we tried to adapt the indicator, and describe an activity that children could recognise. In the 2nd translation, the question was formulated "Debattera och ifrågasätta lösningar i matematik" (Debate and question mathematical solutions). This was made to improve the comprehension, by providing an example Swedish learners would comprehend, but which would still be an indicator of students valuing openness and exploration.

Secondly, we introduced a similar format for all items. We checked that each item consisted of a verb together with an object; for example, the item "Investigations" became in the Swedish version formulated as "Making investigations" (for further methodological discussions on the translation and adaptation process see Andersson and Österling 2013, 2014).

Survey Sample and Data Collection

Our aim was to achieve a spread of students from public as well as private schools, rural as well as urban areas, and a geographical distribution over Sweden. We needed to find mathematics teachers who would be interested in participating, in order to get access to students as respondents. We did not have access to a database of students nor mathematics teachers to be able to plan a random probability sample. Instead, we wanted to achieve a quota sampling, with equal numbers of the both age groups, boys and girls and a geographic spread. To get access to students, we used a convenience sampling through contacting mathematics teachers who previously had been participating in a national mathematics project. These geographically well spread teachers helped us distribute the survey. We analysed respondents to check the relative proportions between the years five and eight students to be approximately equal, and for the distribution of gender, and we achieved a fair distribution.

The participating teachers received a letter of instruction with information regarding the Swedish ethical guidelines (Swedish Research Council 2011) together with the web link. The web-survey was distributed and collected by *Survey and Report* software. We received 850 completed survey forms. Before beginning the statistical analysis, we removed respondents with more than 10% answers missing, which left 742 students' responses.

Analysing Democratic Actions Through Values Behind Survey Items

To be able to tease out the relationship between the students' appreciated activities and hence valuing in mathematics classrooms and the values stated as democratic actions in the national steering documents in mathematics, we turn to the enculturation of mathematical values in schools (Bishop 1988; Seah 2013), which are briefly described in Table.[2]

In each of the three dimensions, a pair of opposing values are described. This can be understood as two opposing positions, and in a certain community or mathematics classroom, values are understood to be positioned somewhere on a continuum between the extreme values. In the ideological dimension, rationalism allows for students to communicate and argue for a solution or line of reasoning in mathematics. This reason with the idea of participation as an agentic activity, and ideas about sharing and understanding the arguments of others. The value of objectism instead focuses application of pre-determined formulae and praxis of symbolising.

Table: Mathematical values (see Bishop 1988)

Dimension	Pairs of opposing values	
Ideological values: the ideology of mathematics	*Rationalism*—reasoning and argument is valued	*Objectism*—symbolising and applying ideas of mathematics is valued
Sociological values: who can do mathematics	*Openness*—mathematics is democratically open for anyone to use and explain	*Mystery*—the mystique of mathematical ideas and their origin and who possesses the power to explore
Sentimental values: What sensations mathematics can bring	*Control*—a sense of certainty and power through mastery of rules is valued	*Progress*—the sense of ideas growing through questioning is valued

[2]Bishop (1988) also cautioned that these six values are discussed in the context only of Western mathematics classrooms.

Table: Items associated with values of openness, progress or rationalism

Value	Questionnaire item
Openness	Small-group discussions
	Whole-class discussion
	Debates
	Explaining where formulae or rules came from
Progress	Relating maths to other subjects
	Stories about recent developments in maths
	Relationships between concepts
Rationalism	Abstract or theoretical aspects of maths
	Learning the proofs
	Verifying theorems or hypotheses
	Learning the proofs
	Verifying theorems or hypotheses

In the sociological dimension, openness is the idea of mathematics being open for inquiry and explanations, thus a description of activities that invite students to actively participate. In activities related to mystery, mathematical ideas are usually already outlined and described by someone else (c.f. Andersson and Wagner 2018). In the sentimental dimension, progress relates to activities where students have the sense of having the opportunity to develop own ideas through questioning and exploring. Control instead describes activities that reinforce the sense of correctness, mastery and certainty (Bishop 1988, 2002).

The most evident values to be connected to democratic actions are found in the sociological dimension, where *openness*, described as mathematics being open to everybody to explain, reasons with our description of democratic participation as *having opportunity to actively contribute to mathematical work on equal terms.*

In the other dimensions, *progress*, allowing ideas to grow through questioning rhymes with the description of democratic participation as *exploring, collaborating and communicating*. Also, the value of reasoning aligns with the ideas of actively *communicating mathematical ideas.* Therefore, in this chapter, we argue that the valuing of rationalism, openness and progress are strongly related to democratic actions in mathematics classrooms (see also Seah et al. 2016). The WiFi-questionnaire had associated research guidelines which suggested how the items in the questionnaire were distributed among mathematical and mathematics educational values. In those guidelines, the items associated with *openness, progress and rationalism* are shown in Table.

The items above are examples of activities in mathematics classroom. Thus, this distribution was our starting point for investigating which democratic actions students valued as important in Swedish mathematics classrooms.

Methods of Statistical Analysis

We conducted a descriptive analysis in SPSS where we calculated means and standard deviations with the aims to find what students value as important or not important among the 64 items in the survey: 1 corresponded to the valuing of *absolutely important*, 5 corresponded to *absolutely unimportant*, and 3 corresponded to a *neutral alternative*.

The initial plan was to use a principal component analysis (PCA) to see whether such components could be interpreted in terms of mathematical values. However, despite different analytical approaches, the resulting components would consist of one large component containing more than a third of the items, and the remaining components did not relate to values. Instead we saw how they consisted of similar activities, as valuing ICT in different forms. Thus, we could not find the hypothesized distribution of research items related to values in accordance with the research guidelines, as described in the previous section. In addition, since our sample was a non-probability sample, we did not have the statistical means for estimating errors. Despite these shortcomings, we did see some patterns in the responses.

As described above, in a classroom aiming for democratic actions; rationalism, openness and progress would be the desired values. Being cautions not to extend beyond the reliability of the sampling or analytical methods, we will present the descriptive statistics, and compare the most and least valued items.

Results

In this results section, we first present the means and standard deviations of all items, sorted from the most valued to the least valued Table. This result is used to describe the importance students attribute to items connected to democracy, in relation to the other items. Thereafter, we focus the most and least valued items, to see what it is students do value.

Results for All Items

Table shows the means and standard deviations for all 64 items.

It will be noticed first; the means indicate that no item is valued as not important, including items related to democratic actions. Second, the standard deviations demonstrate that the spread is small among the most valued items and larger among the least valued items. A larger deviation for items with means at the middle of the Likert-scale is not unexpected, since it deviates to both sides.

Table: Descriptive statistics of responses to the WIFI-questionnaire (N is the number of returns per item)

Items ranked from most valued	N	Mean	Std. Dev. (σ)
1. Explaining by the teacher	742	1.33	0.700
2. Knowing the times tables	742	1.43	0.767
3. Understanding why my solution is incorrect or correct	736	1.45	0.708
4. Learning through mistakes	743	1.54	0.781
5. Remembering the work we have done	737	1.57	0.805
6. Understanding concepts/processes	738	1.57	0.756
7. Alternative solutions	739	1.59	0.841
8. Examples to help me understand	743	1.64	0.778
9. Teacher helping me individually	745	1.69	0.818
10. Knowing the steps of the solution	736	1.70	0.799
11. Looking for different ways to find the answer	741	1.70	0.805
12. Knowing which formula to use	732	1.70	0.797
13. Working step-by-step	740	1.71	0.841
14. Memorising facts	741	1.72	0.849
15. Me asking questions	738	1.73	0.870
16. Problem-solving	736	1.74	0.793
17. Completing mathematics work	737	1.75	0.843
18. Given a formula to use	741	1.75	0.843
19. Shortcuts to solving a problem	737	1.78	0.876
20. Feedback from my teacher	741	1.79	0.865
21. Practicing how to use maths formulae	742	1.79	0.901
22. Connecting maths to real life	740	1.81	0.868
23. Verifying theorems and hypotheses	742	1.83	0.841
24. Knowing the theoretical aspects of mathematics	734	1.85	0.876
25. Getting the right answer	737	1.86	1.006
26. Writing the solutions step-by-step	743	1.86	0.871
27. Mathematics tests/examinations	739	1.89	0.997
28. Looking for different possible answers	744	1.90	0.820
29. Teacher asking us questions	740	1.96	0.912
30. Mystery of maths	732	1.96	0.904
31. Relationships between maths concepts	742	1.96	0.795
32. Using mathematical words	742	1.97	0.937
33. Learning the proofs	736	1.98	0.938
34. Practicing with lots of questions	741	1.99	0.920

(continued)

Table (continued)

Items ranked from most valued	N	Mean	Std. Dev. (σ)
35. Working on the maths by myself	737	1.99	0.912
36. Mathematics puzzles	738	2.00	0.924
37. Hands-on activities	739	2.01	0.961
38. Looking for maths in real life	743	2.06	0.919
39. Mathematics debates	736	2.07	0.922
40. Using diagrams to understand maths	740	2.08	0.870
41. Using concrete materials to understand mathematics	735	2.09	0.868
42. Teacher use of keywords	739	2.11	0.934
43. Explaining where rules/formulae came from	738	2.12	0.940
44. Stories about recent developments in mathematics	742	2.16	0.997
45. Doing a lot of mathematics work	738	2.18	1.022
46. Mathematics homework	736	2.20	1.242
47. Relating mathematics to other subjects in school	740	2.22	0.980
48. Investigations	737	2.22	0.865
49. Small-group discussion	738	2.26	0.919
50. Whole-class discussion	737	2.29	0.979
51. Mathematics games	741	2.29	1.028
52. Explaining my solution to the class	737	2.33	1.146
53. Making up my own math questions	738	2.39	1.046
54. Feedback from my friends	741	2.41	1.089
55. Students posing maths problems	741	2.45	0.935
56. Using the calculator to check the answer	741	2.46	1.084
57. Learning with the internet	741	2.47	1.062
58. Learning with the computer	741	2.51	1.056
59. Outdoor mathematics activities	743	2.52	1.157
60. Using the calculator	738	2.54	1.059
61. Appreciating the beauty of maths	744	2.54	1.059
62. Stories about mathematics	737	2.64	1.128
63. Stories about mathematicians	737	2.74	1.137
64. Being lucky at getting the correct answer	732	2.75	1.212

Table Results related to openness, progress or rationalism

Value	Item	Rank	Mean	Std
Openness	Small-group discussions	49	2.26	0.919
	Whole-class discussion	50	2.29	0.979
	Debates	39	2.07	0.922
	Explaining where formulae or rules came from	43	2.12	0.940
Progress	Relating maths to other subjects	47	2.22	0.980
	Stories about recent developments in maths	44	2.16	0.997
	Relationships between concepts	31	1.96	0.795
Rationalism	Knowing the theoretical aspects of maths	24	1.85	0.876
	Learning the proofs	33	1.98	0.938
	Verifying theorems or hypotheses	23	1.83	0.841

Items Associated with Democratic Actions

From Table, we assemble the items associated with democratic actions through the valuing of openness, progress and rationalism and show the results in Table.

From this list of results, we can conclude that the items related to democratic actions are neither the most, or the least valued activities by the students. Among the three value dimensions, rationalism seems valued as slightly more important compared to progress or openness.

The Most or Least Valued Activities

The distribution of responses for the most valued items are shown in Fig, together with the least valued items.

The three most valued activities were *Explaining by the teacher, Knowing the times tables* and *Understanding why my solution is incorrect or correct*. Explanations by the teacher can be understood as trusting authorities for providing knowledge, but it can also be understood as a righteous demand for access to knowledge and understanding. The other two both refers to correctness, where the third item also stresses the understanding of *why*. We noted a strong similarity between the responses from all the 743 students, in other words both year 5 and the year 8 students responded

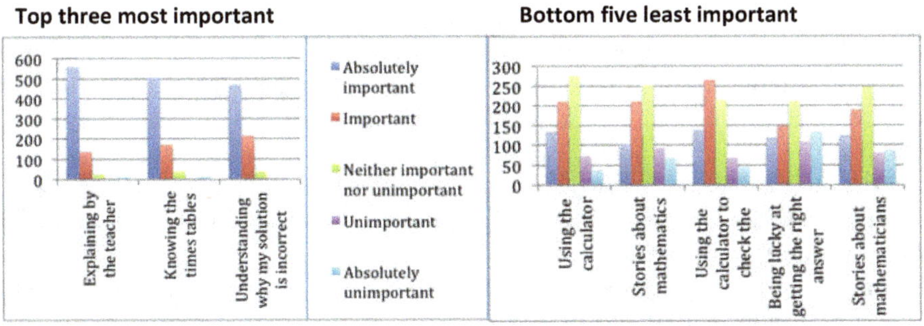

Fig. The diagram displays the number of students (among the 742 respondents) responding to each alternative on top three most important items and bottom five least important items

in the same way and valued the same activities in the top three group. Only 16 respondents in the whole cohort marked "knowing the times tables" as less important.

Among the least valued items, there is a large spread in students' responses. So, rather than being valued as not important, we observed a large individual variance in the importance students attribute to those items. Being at the bottom of the list of valued items here means that an item is still important for many students, however, due to the large spread, we cannot see these items as representing shared valuing. The stories about mathematics or mathematicians were mostly neutrally valued, and might not be familiar for all students. Several students in the pilot test would express that luck has nothing to do with mathematics, this was the one question they would react to spontaneously. Activities including use of the calculator were also to a large extent neutrally valued.

Summing up, the most valued items reflects a valuing of a correct explanation/knowledge or understanding, whereas the least valued items has to do with the calculator or the history of mathematics. In addition, most students find that being lucky at getting the right answer has nothing to do with learning mathematics. We notice that the most valued items are quite common in Swedish mathematics classrooms, whereas the least valued are not.

Discussion

At the departure of this project we were hoping to learn about possibilities to align mathematics teaching with students' valuing of activities for learning, in line with the curricular intentions of democratic participation and actions. However, we arrived at the somewhat discouraging result, that students' valuing of "knowing the times tables" and "teachers' explanations" and "correctness" over explorative, communicational and collaborative activities. We asked ourselves what might be the cultural and historical reasons behind these results.

Mathematics teaching does not exist in a vacuum. It is affected by, and at the same time affects cultural expressions, within the mathematics classroom as well as outside. For example, the Swedish children books and films about Pippi Longstocking, may serve as an illustrative example. Pippi, an orphaned eleven-years old strong, kind but also a stubborn and questioning girl, lives on her own with a horse and a monkey. In one episode, her two very well-behaving friends Tommy and Annika tell her that she needs to attend school. This scene describes the *very first and only* time Pippi ever enters a classroom:

"Hey, everybody," hollered Pippi, swinging her big hat. "Am I in time for pluttification?" (Lindgren 1945, 2007, p. 60).

'Pluttification' tables, or the multiplication tables, as a properly fostered Swedish student knows to name them, are in our results one of the most important activities in mathematics classrooms as seen by students. Even though Pippi is a children's book fiction she captures well the importance assigned to the times tables, as that is the only thing she knows about school. This fiction is still relevant for Swedish children, who share the valuing of times tables as an important part of school.

Lundin (2008) reminds us that when schooling in Sweden became public and mandatory, teachers had to deal with a larger number of first generation children attending school. The first mathematics textbooks, published in the early 1940s aimed to both support mathematics learning and to support teachers in coping with disciplinary problems. "This need led to the promotion of schoolbooks filled with a large number of relatively simple mathematical problems, arranged in such a way that they (ideally) could keep any student, regardless of ability, busy—and thus quiet—for any time span necessary" (p. 376). The teaching at that time hence became a medium for both mathematics learning and fostering children. We remind ourselves that this is the kind of enculturated teaching practice todays students' parents and grandparents experienced. We do not see our results as a reflection of the most common teaching, nor as the most important means for learning mathematics. Instead, we must understand the valuing of times tables or teachers' explanations as an expression of enculturated and therefore culturally valued actions in mathematics classrooms, where this enculturation takes place not only in school, but in conversations with parents, grandparents, in media and in children's books.

At a policy level, different actions are valued. Two recent and important School Inspectorate's research reports concerning primary and secondary school education respectively (School Inspectorate 2009, 2010) concluded that teachers to a large extent rely on textbooks when planning their teaching, hence they trust the textbooks to fulfil curricular objectives. The Inspectorate highlighted the fact that students' individual work still dominates mathematics lessons, thus resulting in mainly mechanical calculations with less time for students' discussions, collaborations and problem solving. Teaching seems to result in under-stimulated students, who experience mathematics as a boring, tedious and sometimes even "stupidizing" [fördummande] subject (School Inspectorate 2010, p. 8). To take this argument one step further, we note that such "stupidizing" activities do not rhyme with the Swedish educational values, based on the declaration of Human Rights (United Nations 1948), stating humans as endowed with reasoning, and that this reasoning should be protected by democracy, and at the same time generate democracy through democratic actions.

Researchers and School inspections has been looking at the rationale or consequences from this particular way of organising mathematics education, with a teaching mainly based on individual student (textbook) work. It is still found to support teachers in managing non-homogeneous student groups, however now with the argument that each student shall be able to work according to his/her previous learning and needs (Johansson 2006). The argument is that this is how democratic participation and inclusion is fulfilled (Hansson 2010; Lingefjärd and Meier 2010). Lundin's (2008) explanation of the historical development resonates with the phenomena described by the School Inspectorate (2009, 2010) although the reasons are different. We add that the particular learning activities; as knowing the times tables and teacher's explanations, are the most valued activities by students. And they were clearly conflicting with the intentions formulated in the steering documents. It is precisely here we find a conflicting valuing and maybe one possible explanation to why the desired transformation of mathematics teaching takes time (Seah et al. 2016). Following this argument, we now discuss how the survey results may be used to better understand why teaching transformation seems so difficult in the mathematics classroom.

In our results, students valued "teacher explaining" as important. We may understand this response as students appreciating good explanations or scaffolding by teachers, or even a special relationship with the teacher. However, it also recognises that the mathematical learning activities are the responsibility of the teachers. This may cause valuing-conflicts, when teachers apply intentions in the curriculum, stating that teachers' starting point shall be "ensuring that all pupils can exercise real influence over working methods, forms and contents" and that "pupils are able and willing to take personal responsibility for their learning" (Skolverket 2011, p.17). Students' valuing of "teacher explaining" reflects an understanding of the learning of mathematics as the responsibility of teachers, rather than as a result of "joint labour" (Radford and Roth 2011) or as an individual democratic responsibility.

The political aim to allow space for students' influence on the planning and evaluation of mathematics teaching, can be expected to align well with the mathematical values openness, rationality and progress where perhaps openness is the most important of the three. Thus, to what extent do mathematics lend itself to such negotiation? Very little, according to Wagner and Herbel-Eisenman (2013, p. 483). They demonstrate the very central authoritative nature of mathematics through its "interest in certainty". In this way mathematics is epistemologically different from other subjects in school, since deductive reasoning based on already stated axioms, rather than empirical explanations and students' own initiatives, are valued. The survey results indicate precisely that students value control through certainty and the mastery of rules. Thereby, arriving at students' influence on their learning and planning of mathematical activities may be specifically challenging in mathematics.

We argue that the contradiction between the political expectations of democratic participation and actions, and the invitation to students to influence teaching on the one hand, and on the other hand students use of this influence through valuing *teacher explaining*, *mastering times tables* and *understanding why the answer is incorrect*, rather conserve a mathematics teaching organised around values as objectism and control than through openness and rationalism. Thus, giving back the influence to the historically supported way of learning mathematics, to the teacher who passes on to the textbooks.

This may explain the dilemma teachers face when opening up spaces for students to influence the classroom work. Aligning teaching with what Swedish students' value involves a risk to conserve a traditional way of mathematics teaching, or in Skolverket's (2011) words, an "exercise learning paradigm". This result also highlights that implementing a more democratic mathematics teaching and applying open learning activities that challenges students valuing may be a long process, where consistent values need to be negotiated and addressed throughout the process: from the planning to the assessment of mathematics knowledge in line with the student Zissy's comment above.

Acknowledgements We are grateful for critical conversations with Prof. Em. Alan Bishop, Prof. Em. Philip Clarkson and Assoc. Prof. W. T. Seah and with dear colleagues at Stockholm University: Anette de Ron and Elisabeth Hector.

References

Aguilar, M. S., & Zavaleta, J. G. M. (2012). On the links between mathematics education and democracy: A literature review. *Pythagoras, 33*(2).

Alvestad, M. & Samuelsson, I. (1999). A comparison of the national preschool curricula in Norway and Sweden. *Early Childhood Research and Practice, 1*(2). Retrieved from http://ecrp.uiuc.edu/v1n2/alvestad.html.

Andersson, A. (2011). A "Curling Teacher" in Mathematics Education: Teacher identities and pedagogy development. *Mathematics Education Research Journal, 23*(4), 437–454.

Andersson, A., & Österling, L. (2013). Measuring immeasurable values. In A. M. Lindmeier, & A. Heinze (Eds.), *Proceedings of the 37th Conference of the International Group for the Psychology of Mathematics Education* (Vol. 2, pp. 17–24). Kiel: PME.

Andersson, A., & Österling, L. (2014). Metric equivalence in international surveys: Cultural edges. In Liljedahl, P., Nicol, C., Oesterle, S., & Allan, D. (Eds.). *Proceedings of the Joint Meeting of PME 38 and PME-NA 36* (Vol. 1). Vancouver, Canada: PME.

Andersson, A., & Valero, P. (2015). Negotiating critical pedagogical discourses: Contexts, mathematics and agency. In P. Ernest & B. Sriraman (Eds.), *Critical mathematics education: Theory, praxis and reality* (pp. 203–228). Charlotte, USA: Information Age Publishing.

Andersson, A., & Wagner, D. (2017). Love and bullying in mathematical conversations. In Chronaki, A. (Ed.). *Proceedings of the 9th International Mathematics Education and Society Conference* (pp. 382–392). Volos, Greece: University of Thessaly Press.

Andersson, A., & Wagner, D. (2018). Re-mythologizing mystery in mathematics: Teaching for open landscapes versus concealment. *Education Sciences, 8*, 41; https://doi.org/10.3390.

Baba, T., Iwasaki, H., Ueda, A., & Date, F. (2012). Values in Japanese mathematics education: Their historical development. *ZDM Mathematics Education, 44*(1), 21–32.

Biesta, G. J. J. (2007). Education and the democratic person: Towards a political conception of democratic education. *Teachers College Record, 109*(3), 740–769.

Biesta, G. J. J. (2009). Good education in an age of measurement: on the need to reconnect with the question of purpose in education. *Educational Assessment, Evaluation and Accountability, 21*(1), 33–46.

Bishop, A. (1988). *Mathematical enculturation: A cultural perspective on mathematics education.* Dordrecht: Kluwer Academic Publishers.

Bishop, A. (2002). Mathematical acculturation, cultural conflicts, and transition. In G. De Abreau Bishop, A., & N. Presmeg (Eds.), *Transitions between contexts of mathematical practices* (pp. 193–212). Netherlands: Springer.

DeJarnette, A. F., & González, G. (2015). Positioning during group work on a novel task in Algebra II. *Journal For Research In Mathematics Education, 46*(4), 378–422.

Einarsdottir, J., Purola, A. M., Johansson, E. M., Broström, S., & Emilson, A. (2015). Democracy, caring and competence: Values perspectives in ECEC curricula in the Nordic countries. *International Journal of Early Years Education, 23*(1), 97–114.

Esmonde, I. (2009). Mathematics learning in groups: Analyzing equity in two cooperative activity structures. *The Journal of the Learning Sciences, 18*(2), 247–284.

Hansson, Å. (2010). Instructional responsibility in mathematics education: modelling classroom teaching using Swedish data. *Educational Studies in Mathematics, 75*(2), 171–189.

Hofstede, G., Hofstede, G. J., & Minkov, M. (2010). *Cultures and organizations: software of the mind: Intercultural cooperation and its importance for survival* (3rd ed.). New York: McGraw-Hill.

Johansson, M. (2006). *Teaching mathematics with textbooks: A classroom and curricular perspective. Doctoral Thesis*. Luleå: Luleå tekniska universitet.

Kurth, L. A., Anderson, C. W., & Palincsar, A. S. (2002). The case of Carla: Dilemmas of helping all students to understand science. *Science Education, 86*(3), 287–313.

Lindgren, A. (2007). *Pippi Longstocking*. London: Penguin Books Limited.

Lingefjard, T., & Meier, S. (2010). Teachers as managers of the modelling process. *Mathematics Education Research Journal, 22*(2), 92–107.

Lundin, S. (2008). *Skolans matematik: en kritisk analys av den svenska skolmatematikens förhistoria, uppkomst och utveckling.* Doctoral Thesis. Uppsala: Acta Universitatis Upsalensis.

Ministry of Education [Utbildningsdepartementet] (2011). *SFS2010:800, Skollagen, 1 kap §4–5.* Stockholm: Riksdagen.

Norén, E. (2015). Agency and positioning in a multilingual mathematics classroom. *Educational Studies in Mathematics, 89*(2), 167–184.

OECD. (2006). *Starting strong II: Early childhood education and care*. Paris: OECD.

Radford, L., & Roth, W. M. (2011). Intercorporeality and ethical commitment: An activity perspective on classroom interaction. *Educational Studies in Mathematics, 77*(2–3), 227–245.

School Inspectorate. (2009). *Skolinspektionens Kvalitetsgranskning, Undervisningen i matematik—utbildningens innehåll och ändamålsenlighet*. Rapport 2009:5. Stockholm: Frizes.

School Inspectorate. (2010). Olika elever—samma undervisning. *Skolinspektionens erfarenheter och resultat från tillsyn och kvalitetsgranskning 2010*. 2011:4396. Stockholm: Frizes.

Seah, W. T. (2013). Assessing values in mathematics education. In A. M. Lindmeier & A. Heinze (Eds.), *Proceedings of the 37th Conference of the International Group for the Psychology of Mathematics Education* (Vol. 4, pp. 193–200). Kiel, Germany: PME.

Seah, W. T., & Andersson, A. (2015). Valuing diversity in mathematics pedagogy through the volitional nature and alignment of values. In A. Bishop, H. Tan, & T. Barkatsas (Eds.), *Diversity in mathematics education: Towards inclusive practices* (pp. 167–184). Switzerland: Springer.

Seah, W. T., Andersson, A., Bishop, A., & Clarkson, P. (2016). What would the mathematics curriculum look like if values were the focus? *For the Learning of Mathematics, 36,* 1.

Skovsmose, O. (2005). *Travelling through education: Uncertainty, mathematics, responsibility*. Rotterdam: Sense Publishers.

Skolverket, (2011). *Curriculum for the compulsory school, preschool class and the recreation center*. Stockholm: Skolverket.

Stemhagen, K. (2016). Deweyan democratic agency and school math: Beyond constructivism and critique. *Educational Theory, 66*(1–2), 95–109.

Survey Research Center. (2010). *Guidelines for best practice in cross-cultural surveys.* In J. Harkness (Ed.). Retrieved January 09, 2013, from Institute for Social Research, University of Michigan, A. Arbour. http://www.ccsg.isr.umich.edu/.

Swedish Research Council [Vetenskapsrådet] (2011). *Good research practice.* [God forskningssed]. Stockholm: Vetenskapsrådet.

Swan, M. (2014). Improving the alignment between values, principles and classroom realities. In Y. Li & G. Lappan (Eds.), *Mathematics curriculum in school education* (pp. 621–636). Netherlands: Springer.

United Nations. (1948). *General Assembly Resolution 217 (III).* Paris: United Nations Official Documents.

United Nations (2016). *Global issues. Democracy.* Retrieved September 25 2016 from http://www.un.org/en/globalissues/democracy/.

Wagner, D., & Herbel-Eisenmann, B. (2009). Re-mythologizing mathematics through attention to classroom positioning. *Educational Studies in Mathematics, 72*(1), 1–15.

Wagner, D., & Herbel-Eisenman, B. (2013). Disbursing authority among mathematics students. *Proceedings of the Seventh International Mathematics Education and Society Conference* (pp. 483–492). Cape Town: MES7.

Wagner, D., & Herbel-Eisenmann, B. (2014). Identifying authority structures in mathematics classroom discourse: A case of a teacher's early experience in a new context. *ZDM: The International Journal of Mathematics Education*, *46*(6), 871–882.

Wood, M. B. (2013). Mathematical micro-identities: Moment-to-moment positioning and learning in a fourth-grade classroom. *Journal for Research in Mathematics Education, 44*(5), 775–808.

13

The Elementary Mathematics Teachers' Values Underlying Teacher Noticing: The Context of Polygons

Fatma Nur Aktaş, Esra Selcen Yakıcı-Topbaş and Yüksel Dede

Abstract Teachers are expected to make decisions that respond to the needs of students in classroom practices. Teacher noticing emphasizes that teachers should decide how to respond to situations in classroom practices. Moreover, one of the variables that influence teachers' decision-making skills is teachers' values. The purpose of this chapter is to examine elementary teacher values in terms of the decision making process underlying noticing in a specific mathematical domain, namely polygons. We have conducted this qualitative study, designed as a case study, with five elementary mathematics teachers working at elementary schools in Turkey. The participants were selected using convenience sampling. The data were collected with video-recordings of classrooms and semi-structured interviews and were coded using content analysis approach. Teachers' values were presented in the context of teacher noticing, which is a situation-specific skill. The results shed light on the relationship between teachers' values and teacher noticing, which focused the decision-making perspective.

Keywords Teachers' values · Teacher noticing · Decision-making · Elementary mathematics teachers · Polygons

---

F. N. Aktaş (✉)
Division of Mathematics Education, Gazi Education Faculty, Department of Mathematics and Science Education, Gazi University, Ankara, Turkey
e-mail: fnuraktas@gmail.com; fnuraktas@gazi.edu.tr

E. S. Yakıcı-Topbaş · Y. Dede
Gazi University, Ankara, Turkey
e-mail: selcenyakici@gazi.edu.tr

Y. Dede
e-mail: ydede@gazi.edu.tr

Teacher Noticing, Decision-Making, and Teacher Values

In designing effective teaching environments, classroom culture may be listed alongside the components of teaching dynamics like curriculum, lesson plan, teacher, and student. Among the variables of classroom culture that are important for effective learning are the importance of student thinking, student engagement, coordinating the classroom environment, and the classroom setting, all of which have been emphasized in the literature (Husband 1947; Muijs and Reynolds 2010, pp. 2–3). When it comes to mathematics teaching/learning, The National Council of Teachers of Mathematics [NCTM] (2000) expects teachers to not only teach mathematical knowledge, but have a good understanding of what students know and what kind of support for their studying they need, in order for them to learn better. Weissglass (2002, p. 35) recommended an even wider perspective that includes a school's culture and role, as well as mathematical knowledge, learning, and instruction.

When considering such frameworks as these, various roles are clearly expected of teachers for effective teaching, such as managing "the blooming, buzzing confusion of sensory data" (Sherin and Star 2011, p. 69) that they encounter while teaching. Within this 'confusion' teachers need to determine which elements to register as important and which to ignore, depending on the learning environment. Therefore, which types of circumstances teachers pay attention to, how they interpret these circumstances, and their processes for implementing these interpretations through teaching activities are significant. The combination of these processes can be defined as teacher noticing (Sherin et al. 2011; van Es and Sherin 2002).

Specifically, 'noticing' by the teacher in the classroom can be described as seeing and understanding particular aspects of classroom instruction. While this concept is static in some professions, it is active for teachers as it depends on the interplay between the classroom culture and environment, the curriculum and content, and the students and teacher. The structure of noticing concentrates on the elements that teachers pay attention to; that is, noticing is related to what they choose to respond to or ignore, as well as how they interpret these elements within the intricacy of the classroom.

Various studies exist in the literature on conceptualizing teacher noticing (see, Sherin et al. 2011). In the current study, teacher noticing is situated within a framework of teachers attending to the features of classroom interaction that occur during instruction (such as students' misconceptions, confusion, concurrent requests to speak, and so on), their reasoning and interpreting of what they did attend to, and how the teachers used these reasons and interpretations to decide why and how to respond in the ongoing teaching processes (Jacobs et al. 2010; van Es and Sherin 2002). It is the last component of this sequence, deciding how to respond (Jacobs et al. 2010), which became an important element in this study. This element is observable in the teachers' active instructional processes, whereas the first two elements cannot be readily observed in the ongoing teaching/learning process.

Teachers have to make decisions within each of the three components of the above process. Such decisions will take into account the sum of physical and cognitive efforts related to selecting and preferring various conditions (Taşçı 2011). In particular decision-making regarding why to respond and then what to do in responding (the second and third components of the above process), is the teacher making such decisions all 'made in the moment' when teaching mathematics. But teacher noticing, the first component of the above process, is also influenced by teachers' decision-making processes since they are 'deciding' what to attend to and what to ignore.

Teacher noticing and related decision making have been examined in the past. It has been suggested that teachers' decisions are impacted by school culture, the teacher's mathematical knowledge and their knowledge of mathematics teaching, the teaching situation, goals, learning outcomes, teacher's experience, and their beliefs and values (Bishop and Whitfield 1972; Jacobs et al. 2010). Investigations have also related decision-making processes in terms of teachers' internal knowledge, beliefs, attitudes, and practices (Bartlett 1932; Fisher et al. 2014; Vondrová and Žalská 2012). Dede (2013) inferred decision-making to be basically influenced by values.

The purpose of this chapter is to explicitly examine teachers' values related to the decision making process in the moment of teachers' noticing in mathematics lessons. It is a contention of this chapter that teacher values are principal factors in the moments when teachers make their choices and decisions (Aktaş and Argün 2018; Dede 2013).

Method and Procedure

The Study Design

The study has been designed as a qualitative case study involving five elementary mathematics teachers in Turkey. The seventh-grade teachers, working in three different elementary schools, were selected through convenience sampling. Each participant is taken as a single case; hence, this study was designed as a multiple case study (Yin 1994). The cases are named T1, T2, T3, T4, and T5. T1 and T4 are females. Teachers' teaching experience is 1, 13, 14, 23, and 41 years, respectively. The three schools are located in the same geographical and socio-economic district. T1 works at a small school with only one seventh-grade class. The school has a small number of students and focuses on student learning apart from exams. T2, T3 and T5 work at the same school, which has 20 groups of seventh graders. Because this school considers the high school entrance examination to be important, examination preparation tests frequently take place. Although these three participants work at the same institution, their focus in their lessons varies greatly. While T2 stays self-focused during a lesson, T3 is far more focused on student achievement. T2 utilizes a different resource book to T1 and T5 and presents diverse questions. T3 in contrast designs her lessons by simultaneously considering conceptual and operational comprehension. T5 pays attention to the use of mathematical language and student achievement in her teacher-centered classes. T4's school has four classes of seventh graders, with differences in culture in the different classrooms. The school follows the curriculum strictly. T4 designs her classes by relying on the relevant resource textbook.

Teachers' values (Bishop 2008a, pp. 191–203) and teachers' noticing (Vondrová and Žalská 2012) depend, at least to some extent, on the lesson content. In this study, the notion of regular polygons was the focus for all class groups. In general geometry, questions are often visual questions that demand both careful reading and spatial abilities that include drawing skills, as well as algebraic expressions later on. Therefore, the questions can give rise to critical issues for mathematical communication, which applies no less to the specific field of polygons. Thus, this whole field can provide rich data for extending the knowledge about teacher noticing.

Procedure

The data were collected using video-recordings from the classrooms followed by semi-structured interviews with the teachers. Four of each teacher's classes, each lasting 45 min, were recorded while the participants taught the concept of polygons. Next, the researchers identified the key points in the recordings when a teacher attended and responded to, or ignored, specific situations. For example, such a situation occurred when a teacher was asking a question, he recognized that a student had raised their hand, and the teacher responded to that student's action. But such a situation could be when a teacher failed to recognize those sitting in the back row were not paying attention, or when a student was thinking differently to the way the teacher had expected.

Following the teaching sessions, each teacher was individually interviewed twice, which yielded ten interviews in total. Each interview lasted between 2–3 hours. During the interviews, participants were first asked to note the key points that they had attended to in their lessons. Secondly, the researchers asked them whether they had paid attention to the key points that the researchers had identified in the video-recordings. In both phases, teachers were asked to clarify how and why they had decided to respond or not respond to the various key points for some of the cases where the researchers' had pre-identified key points of the lesson. In this way not only were the participants' confirmation obtained regarding their noticing, but data were also gathered in terms of the values underlying their decision-making.

From this analysis it became clear that the teachers had developed strategies for dealing with decision making at key points of the lessons, as well as being able to assess the effectiveness of those decisions. We were able to identify circumstances

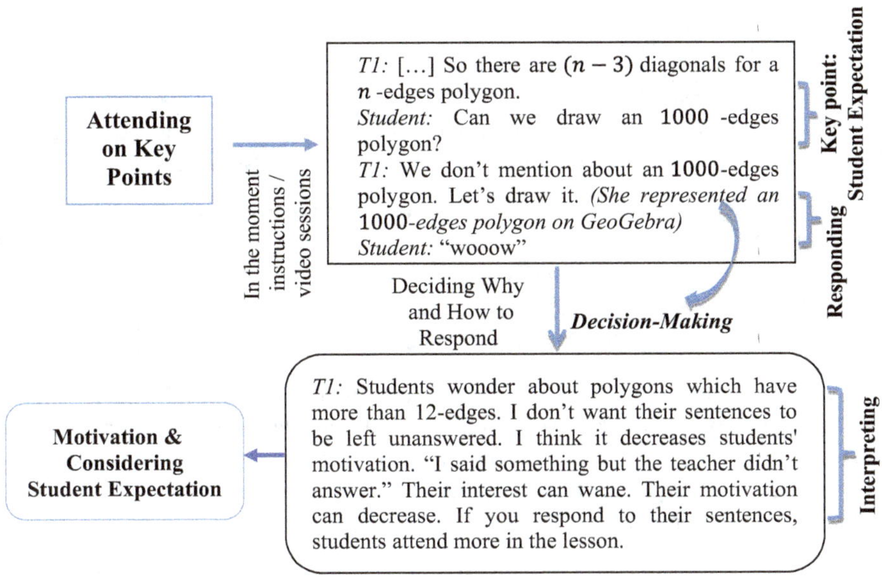

Fig. Analysis process from attending to value

where a teacher used the same strategy at least three times at key points. This then allowed us to reflect on what values, if any, were in play during these similar situations.

In summary then, this study sort to reveal the values underlying the reasons for the five teachers' first noticing and then responding at key points in lessons that were designed to teach the mathematics of polygons. In this chapter '*key point*' refers to both routine and non-routine situations, those that might have been expected by the teacher and those that were unexpected (see Bishop 2008b; Rowland et al. 2015). Content analysis was adopted as the data analysis procedure for the study (Merriam 2009). We interpreted how the teacher responded to these occurrences by interviewing the teacher as we both watched video recordings of each lesson. We then made a determination as to what values lay underneath her/his decision while responding or not responding. An example of the data analysis is summarized in Fig.

After repeated analyses by the researchers a consensus emerged as to various categories to which the teachers' values could be assigned. An 84.7% consensus rate among the researchers was obtained for this process (Miles and Huberman 1994). The following section outlines in more detail the results that were obtained from this process.

Teacher Values

This section discusses the underlying values that were influencing teachers' noticing and their decision making at the key points of lesson, and identifies the categories that subsequently emerged from this analysis. Some sample situations are presented.

> T5: How do we express the situation herein? (The sum of two inner angles is equal to the outer non-adjacent angle)
>
> Student: The sum of the two inner angles equals the outer angle that has the non-shared edge.
>
> T5: Which angle?
>
> Student: The angle not included in the calculations, separate from what's been given.

In this situation, the teacher T5 probes the student's understanding of angles associated with polygons. This key point was examined with T5 in an interview with the video recording available. The point of the interview was to try and understand how he had interpreted the situation and the implicit values underlying this process. The teacher commented:

> T5: [...] I want to know all students' opinions whether correct or incorrect. I want to offer the students that opportunity. Additionally, I want to let them express themselves accurately. The student said "edge" and "length" [at an earlier point] but not "vertice." Angles do not have length. Having them try to find the true answer is necessary for all students, even if it is inaccurate in the moment. This will stay in their mind. In other words, to ensure most of the students participate... They express things one by one, one mentions angle while another says edge, whereas they must be expressed as a whole, clearly and accurately.

T5 was trying to create a classroom atmosphere where students could express themselves in an open-minded fashion, which is a way to create a democratic classroom setting, the first of the values categories we noted (see Fig. later). Within this democratic setting, various values can be found. One is *freedom of expression* and another is *equality,* which in this context relates to all students having the same opportunities and rights to contribute to the ongoing class discussion or in other words a sense of fairness in the students' in-class opportunities.

Another point in T5's expressions is that of using *mathematical language*. Using *mathematical language* can be considered a prerequisite for improving mathematical communication skills, including the types of oral and written communication between students with their teachers, as well as among each other. Bishop (1991, pp. 69–72) noted the importance of mathematical language with all its variety: the use of mathematical symbols and not just words; how to express results in an appropriate manner at the conclusion of a problem solving process; and how students should be encouraged to create and utilize models and diagrams. In this study the use of *mathematical language* based on the characteristics of the concept of polygons was important.

T5's statement also draws attention to the valuing of *rigorous*. In this instance the teacher was valuing *rigorous* in verbal expressions, an aspect of the more general notion of mathematical language. Mathematical communication skills, both verbal and written, necessitate individuals to be able to express themselves openly and to use mathematical language accurately and effectively (NCTM 2000). Based on this necessity, students' skills at being able to make mathematical definitions with clear, open, and accurate expressions, at explaining operations or solutions, and at expressing their opinions all give an indication of rigor.

Retention stands out as a value underlying teacher decision making in the classroom, and is at times linked to a teacher's focus on achievement. Simpson and Weiner (1989) define retention as the ability to remember things. In this chapter, the definition has been adopted. When the T5 says in the above excerpt, "*This will stay in their mind*" is a clear indicator of T5's emphasizing knowledge *retention*.

By keeping in mind that mathematical knowledge is cumulative, others have noted that *generalizations* are the most visible components of mathematical thinking being reflected in classroom practices (Bishop 1991, pp. 72–75). Various formulas in the concept and application of polygons have an important role in terms of the number of learning outcomes in the Turkish national mathematics curriculum. T5 is clearly dealing with generalizations when the comment is made that "*Angles do not have length*".

Although not demonstrated by the written text, the manner in which T5 taught in the classroom and his fervor, which was evident in the interview, also showed an important relationship between *conviction,* conceptual understanding, and mathematical communication. Investigating the accuracy of mathematical knowledge and being convincing are both important in the mathematics learning process. In addition, teachers have to be convinced of students' answers and solutions. Moreover, students' beliefs in a mathematical expression, representation, or modeling are also necessary for learning. Therefore, teachers utilize mathematical process skills and technology to *convince* students. Teachers' focus on convincing students of the validity and accuracy of a *generalization* based on rational argumentation, is shown in another excerpt from T5:

> T5: [...] So I did it, but why did I do it? Student should be able to answer its reason. So I want to protect from them rote memorization. There is no selfishness here. No dictation for what I say! Everything is for the students.

The emphasis here is on persuading, or a *convincing* value, which stems from the nature of mathematics. This situation is an important finding since T5 was teaching in a multiple-choice exam-oriented school culture.

Another salient values category that emerged in our analysis was *esoteric*. At one identified key point during a lesson, T4 was saying:

> T4: I will give formulas based on n-numbered polygons for the number of a regular n-edge polygon and the measurements of its interior edges. The formula will give us a direct answer.

In discussing this with the researchers during the interview phase, T4 offered the following explanation:

> T4: It is hard for students to comprehend [the formula]. Proving it is difficult for them at this age. According to the students' levels, this will be more challenging. It is not easy to make proofs and provide reasons. It is longitudinal and full of various symbols, which is why it is challenging. Student cannot comprehend it.

Esoteric can be defined as something understood or addressed by a particular group. In this instance the group who understands the knowledge of formulae is the teacher. T4 chooses to just give the formula in this instance since her decision is the students will not comprehend how it is derived. Others may wonder as to whether this decision was a 'good' decision by T4, but the key point was observed during the lesson, and later the explanation of why that decision was made was given by T4 in the interview. It seems that *esoteric* is an appropriate term to use for this value.

Teachers need to take students' pre-knowledge into consideration when deciding to give feedback to students' stimuli, when using repetition in their practices, and when choosing problems and exercises, or in short, when designing the course. Moreover, teachers have been found to take into consideration the relevance of concepts in knowledge construction. This situation, a natural outcome from the nature of mathematics, has been found effective in teachers' decision making as a result of the teacher's focus on achievement. The following excerpt from T1 is an example of this:

> T1: After parallelogram concept, equilateral quadrangle comes next and I wanted to show the differences between them by drawing side by side. For the rest square rectangle, parallelogram and equilateral quadrangle are confused with each other. [...] Their readiness level might not be suitable for it. It will be acquired in time, they have just learnt it [...].

Teachers' problem-solving strategies, such as accuracy and consistency in solving problems, and adhering faithfully to the lesson processes embedded in the plan designed by the teacher, includes controlling the class. In this process, one can say that teachers are focused on the concept, the lesson plan, and time, or in short the product and result. The following dialogue can be given as an example of considering *control* in another dimension:

> T1 draws various polygons.
>
> Student A: What is the number of diagonals for a twenty-sided polygon?
>
> Student B: Number of diagonals drawn from one edge is 9.
>
> T1: (Silent. No response: Without feedback or confirmation, the teacher goes on drawing).

This interaction shows how the teacher used their silence to help students focus on the key conceptual point; the underlying reason was revealed after the interview. In this case, the fact that the teacher has emphasized *freedom of expression* alongside *control* is worth noting. Why *control* and *freedom of expression* were analyzed under different categories in our study will be discussed in the next section of this chapter.

The value of *judgment*, which emphasizes the teacher's class *authority*, is similar to *control*, but there are also differences. *Judgment* includes a teacher questioning students' solutions, answers, opinions, and reasoning, as well as a teacher evaluating their accuracy. While these processes may be desirable for effective learning, what is implied here is that teachers prioritize their own ideas and solution strategies and exercises a judicial provision for which students' provide feedback. T5's general opinions about the in-class role of the teacher showed that he exercised *judgment*. T5's statement is an example of this meaning:

> T5: […] The teacher is the leading actor with an active position, and the students are like the audience, but, I believe they can be more successful if we can save students from the role of audience, include them in the play, and give them a role in it.

This has been interpreted as the teacher strongly wanting to do something such as deciding to teach a concept, or solving a problem, but not by being the sole player in the drama and the students relegated to just listening. In other words, the teacher perceived that their decisive focus on doing something quite different and bringing the students into the drama as players, albeit with different roles to his as the teacher, shows the exercising of *judgment* and of power to some extent. Dede (2013) placed authority in two categories: absolutist and semi-absolutist. What emerged from this study was the teacher's presence as an authority resembling semi-absolutism, situations where students are included in the mathematical process and interestingly direct the teacher's behavior to a certain extent. T5 is the participant who frequently reflected on this value and clearly expressed having adopted this value in the interviews.

During the interview with T1, she noted "*Ezgi (student) knows this but* because *she made a calculation mistake, she reached the wrong conclusion. We must tell the students that they need to be very careful*." This suggests that for T1, *motivation* and *self-confidence* are important in the cognitive dimension. But T1 also did not neglect the affective dimension. During a lesson T1 encouraged a student with *"You can do it Mustafa (student)!"*. Motivation plays an essential role for students' academic achievement in terms of the choice of activities made by the teacher, but also the level of effort, persistence, and emotional reactions displayed by the students. The latter are clearly privileged by the teacher with words of encouragement. Motivation is defined in the literature as "an intrinsic energy or mental power" (Sternberg and Williams 2002, p. 345). Before introducing self-efficacy as a key factor in social cognitive theory, Bandura (1997) had dealt with human motivation regarding outcome expectations. As a value, *motivation* impacts on teachers' decisions to privilege freedom of expression and equality, and is expressed by using supportive gestures, and giving verbal feedback to students regarding their ideas or thoughts.

Self-confidence is the judgment where an individual feels one's self to be valuable (Bandura 1997, p. 11). Stipek et al. (1998) stated that, while the teacher's objective is for students to understand and learn concepts using *motivation*, motivation inspires in a way that will raise students' willingness to solve a problem and increase positive ambitions through *self-confidence*.

Mathematical reasoning plays a key role as a means in individuals' communication and connection processes for mathematical learning and to be able to use what they learn in daily life. Teachers' lesson designs that aim to develop students' skills of reasoning, expressing and defending their opinions, interpret data obtained from experience, and attempts at making predictions, are all outcomes of reasoning. The situations that have been mentioned, which Bishop (1991) categorized as rationalism, have been addressed as *mathematical reasoning* in this study's findings.

An individual's value system plays a crucial role in one's preferences or choices of which value to privilege in the moment of decision in the classroom (Bishop et al. 2003, pp. 721–725). This notion of choosing between values can be described as a teacher's flexibility. *Flexibility* is defined as the ability to change to suit new conditions or situations (Simpson and Weiner 1989). It also indicates someone who can change their decisions or thoughts easily according to a situation. Teachers do make changes to their lesson plans during the flow of a lesson by taking student expectations, efficiency, or technological variables into consideration. Often they are able to change course easily when meeting unexpected situations by keeping alternative course plans in mind.

Efficiency, an indicator of *flexibility*, is defined in the dictionary as "a good use of time and energy" (Simpson and Weiner 1989). In this context efficiency emphasizes the designing of activities for a lesson, preparing whole lesson plans, and preparing for situations where more goals are reached in a shorter time by keeping possible alternative instructional variables in mind. *Flexibility* also has an element of trying to foresee and considering a variety of students' expectations. This then covers both changes during unexpected situations in classroom practice, as well as pre-planning lessons by considering students' affective, cognitive, and psychomotor statuses. So a teacher showing flexibility would be thinking about student expectations and wishes, considering a range of materials to use and carefully making the problem selection including connecting a concept to daily life, and finding out about students' pre-knowledge of the concepts.

Discussion, Implications and Conclusion

Based on this study's findings, we may state that teacher noticing can be added as a new variable to Bishop and Whitfield's (1972, p. 6) decision-making framework. The values underlying teacher noticing, and discussed in the previous section, are given in Fig.

The teacher values that underlie noticing have been grouped under three categories: *advanced mathematical process, democracy*, and *achievement.* These values are of course influenced by the education and examination system in Turkey and reflect the classroom and school culture. Although we only studied with elementary mathematics teachers, this model can possibly shed light on future studies in order to provide a framework for noticing, values, and the relationship between them, at all levels of school education.

As stated in the mathematics school curriculum and various mathematics education institutions/organizations in various countries (see NCTM 2000; OECD 2013; Taiwan Ministry of Education 2013; Turkish Ministry of National Education [MEB] 2017), mathematical communication is a mathematical process skill that makes mathematical thinking visible in the processes of mathematical comprehension. Utilizing mathematical symbols, terms, and mathematical opinions accurately and effectively and interpreting their accuracy and meaning can be mentioned is an important aspect of the development of mathematical communication skills.

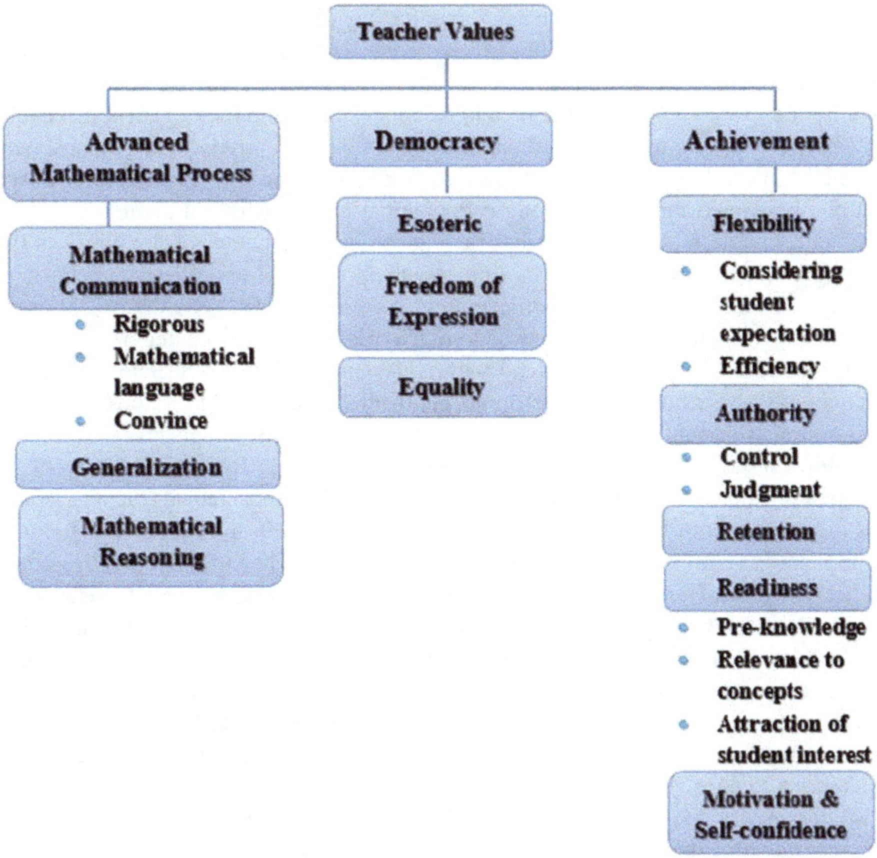

Fig. Teacher values underlying teacher noticing

The indicators of the valuing of rigorous, mathematical language, and convince value lead to the mathematical communication's sub-category of *advanced mathematical processes*. When considering the skills of making connections and communication, the interactions between these becomes important. Taking advantage of reasoning skills is necessary for being able to make mathematical connections, as well as for connecting pre-knowledge to mathematical reasoning. A similar situation also occurs for mathematical communication. While designing a setting for discussion occurs as a communication value in the literature (Seah et al. 2014), this study considers it as an *advanced mathematical process* as this situation creates a setting for reasoning skills. Generalization is also another sub-category fitting under advanced mathematical thinking because in this study it has a dimension that focuses on concepts.

This study has given emphasis to the role of equality in doing/learning mathematics (Seah et al. 2001). Little difference exists between *democracy* and openness.

Openness provides students with a democratic way of expressing their ideas in class (Bishop 1991, pp. 75–77). We have interpreted *democracy* as a value that emphasizes the equality of opportunities in education. Thus democracy was demonstrated in this study when teachers attempted to reach all students in their classroom. A sub-category of esoteric was included under this theme. In this study, when the teachers seemed to be following aims that allowed students freedom to express themselves openly, it seemed they were giving each student a right within a democratic environment. Freedom of expression has been taken as an indicator of flexibility in the literature (Dede 2013). Within this study teachers were observed to pay attention to students' needs and academic achievement. As such, flexibility is the reflection of *achievement* in practice. In short, freedom of expression is a sub-category of *democracy* while flexibility is a sub-category of *achievement*. Similarly, efficiency focuses on *achievement*, which categorically differs from studies in the literature.

Flexibility and authority are quite similar to Dede's (2013) categories. The national transition system of the secondary education examination in Turkey has an impact for this pair of values in this study. School culture noticeably impacted on teachers' noticing in some schools. For example, teacher T4 who worked in an examination-focused school paid attention to this aspect of the school culture when to implementing some of her decisions. But T1, who worked in a student-focused school, reflected far more flexibility in the way she taught in her classroom by being able to give more attention importantly to student feedback, and the level of their engagement. The full impact of the school culture is an issue that should be examined in future studies.

Teachers were found to adopt the values of retention and readiness because they give importance to student achievement. The indicators of student-focused readiness have been categorized as pre-knowledge, relevance to concepts, and attracting student interest. In addition, motivation and self-confidence have been found as other interesting sub-categories. Even though value categorizations could be obtained in the affective dimension, the data has indicated that teachers focus on process and achievement in the cognitive dimension.

References

Aktaş, F. N., & Argün, Z. (2018). Examination of mathematical values in classroom practices: A case study of secondary mathematics teachers. *Education and Science, 43*(193), 121–141.

Bandura, A. (1997). *Self-efficacy, the exercise of control*. New York: W.H. Freeman and Company.

Bartlett, F. (1932). *Remembering: A study in experimental and social psychology*. Cambridge, UK: Cambridge University Press.

Bishop, A. J. (1991). *Mathematical enculturation: A cultural perspective on mathematics education*. Netherlands: Kluwer Academic Publishers.

Bishop, A. J. (2008a). Research, effectiveness, and the practitioners' world. In P. Clarkson & N. Presmeg (Eds.), *Critical issues in mathematics education* (pp. 191–203). New York: Springer.

Bishop, A. J. (2008b). Decision-making, the intervening variable. In P. Clarkson & N. Presmeg (Eds.), *Critical issues in mathematics education* (pp. 29–37). New York: Springer.

Bishop, A. J., & Whitfield, R. (1972). *Situations in teaching*. London: McGraw Hill.

Bishop, A. J., Seah, W. T., & Chin, C. (2003). Values in mathematics teaching the hidden persuaders? In A. J. Bishop, F. Leung, J. Kilpatrick, & K. Clements (Eds.), *Second international handbook of mathematics education* (pp. 717–766). Dordrecht: Kluwer Academic Publisher.

Dede, Y. (2013). Examining the underlying values of Turkish and German mathematics teachers' decision making processes in group studies. *Educational Sciences: Theory and Practice, 13*(1), 690–706.

Fisher, M. H., Schack, E. O., Thomas, J., Jong, C., Eisenhardt, S., Tassell, J., & Yoder, M. (2014). Examining the relationship between preservice elementary teachers' attitudes toward mathematics and professional noticing capacities. In *Research trends in mathematics teacher education* (pp. 219–237). Cham: Springer.

Husband, G. R. (1947). Effective teaching. *The Accounting Review, 22*(4), 411–414.

Jacobs, V. R., Lamb, L. L., & Philipp, R. A. (2010). Professional noticing of children's mathematical thinking. *Journal for Research in Mathematics Education, 41*(2), 169–202.

Merriam, S. B. (2009). *Qualitative research: A guide to design and implementation*. San Francisco: Wiley.

Miles, M. B., & Huberman, A. M. (1994). *An expanded sourcebook qualitative data analysis* (2nd ed.). California: Sage Publications.

Ministry of National Education (MoNE). (2017). Talim ve Terbiye Kurulu Başkanlığı, İlköğretim matematik dersi (5-8.sınıflar) öğretim programı [Board of Education, Elementary school mathematics curriculum (5–8th grades)]. Ankara: MEB Basımevi. (in Turkish).

Muijs, D., & Reynolds, D. (2010). *Effective teaching: Research and practice*. London: Paul Chapman.

National Council of Teachers of Mathematics [NCTM]. (2000). *Principles and standards for school mathematics*. Reston, VA: Author. http://standards.nctm.org/.

Organization for Economic Co-operation and Development [OECD]. (2013). *PISA 2012 assessment and analytical framework: Mathematics, reading, science, problem solving and financial literacy*. Retrieved March 8, 2015, from http://dx.doi.org/10.1787/9789264190511-en.

Rowland, T., Thwaites, A., & Jared, L. (2015). Triggers of contingency in mathematics teaching. *Research in Mathematics Education, 17*(2), 74–91.

Seah, W. T., Bishop, A. J., FitzSimons, G. E., & Clarkson, P. C. (2001). *Exploring issues of control over values teaching in the mathematics classroom*. Paper presented at the 2001 Annual Conference of the Australian Association for Research in Education, Fremantle, Australia.

Seah, W. T., Zhang, Q., Barkatsas, T., Law, H. Y., & Leu, Y. C. (2014). Mathematics learning in mainland China, Hong Kong and Taiwan: The values perspective. In C. Nicol, S. Oesterle, P. Liljedahl, & D. Allan (Eds.), *Proceedings of the Joint Meeting of PME 38 and PME-NA 36* (Vol. 5, pp. 145–152). Vancouver, BC: PME-NA and PME.

Sherin, B., & Star, J. R. (2011). Reflections on the study of teacher noticing. In M. G. Sherin, V. R. Jacobs, & R. A. Philipp (Eds.), *Mathematics teacher noticing: Seeing through teachers' eyes* (pp. 66–78). New York: Routledge.

Sherin, M., Jacobs, V., & Philipp, R. (2011). *Mathematics teacher noticing: Seeing through teachers' eyes*. New York: Routledge.

Simpson, J., & Weiner, E. S. (1989). *Oxford English dictionary online*. Oxford: Clarendon Press. Retrieved July 2017, from http://www.oxfordlearnersdictionaries.com/.

Sternberg, R. J., & Williams, W. M. (2002). *Educational psychology*. Boston: Allyn and Bacon.

Stipek, D., Salmon, J. M., Givvin, K. B., Kazemi, E., Saxe, G., & MacGyvers, V. L. (1998). The value (and convergence) of practices suggested by motivation research and promoted by mathematics education reformers. *Journal for Research in Mathematics Education, 29*(4), 465–488.

Taiwan Ministry of Education. (2013). *Grade 1–9 curriculum guidelines mathematics*. Taipei: Ministry of Education.

Taşçı, Ö. (2011). *İlköğretim öğretmenlerinin okul yöneticilerinin karar verme sürecindeki etkilerine ilişkin algıları ile örgütsel adanmışlıkları arasındaki ilişki*. Yüksek Lisans Tezi, Yeditepe Üniversitesi, Sosyal Bilimler Enstitüsü, İstanbul. (in Turkish).

van Es, E. A., & Sherin, M. G. (2002). Learning to notice: Scaffolding new teachers' interpretations of classroom interactions. *Journal of Technology and Teacher Education, 10,* 571–596.

Vondrová, N., & Žalská, J. (2012). Do student teachers attend to mathematics specific phenomena when observing mathematics teaching on video? *Orbis Scholae, 6*(2), 85–101.

Weissglass, J. (2002). Inequity in mathematics education: Questions for educators. *The Mathematics Educator, 12*(2), 34–39.

Yin, R. K. (1994). *Case study research: Design and methods* (3rd ed., Vol. 5). Applied Social Research Methods Series. London: Sage Publications.

14

Exploring Teachers' Values and Valuing Process in School-Based Lesson Study: A Brunei Darussalam Case Study

Nor Azura Abdullah and Frederick Koon Shing Leung

Abstract In recent years, there is an interest of studies emphasising on the importance of values in mathematics education (Bishop 1999; Macnab 2000; Pa 2009). This study focuses on teachers' values and valuing process within the context of lesson study in school settings. By aligning the lesson study process with Raths et al. (1987) valuing process framework and Bishop's framework of mathematical values of teachers (1988), this study provides a platform for the valuing process to emerge, and this may help to articulate teachers' values in mathematics teaching. This study found that the lesson study processes enabled the value indicators to be observed and studied at three different levels in the curriculum, namely; intended, implemented and attained. By studying the value indicators closely, it was discovered that the teachers' values may get affected by the bilingual context in the mathematics classroom of Brunei, which were also affecting teachers' instructional practices.

Keywords Fractions · Lesson study · Mathematical values · Valuing

Introduction

Researching values in mathematics and mathematics education has increased in recent years. Bishop (2008) proposed that to study the intention behind teachers' classroom practices is perhaps to study them from a values perspective. Bishop suggests this to be done from a socio-cultural perspective, particularly at the pedagogical and individual levels (Bishop 1988). Understanding teachers' values in teaching mathematics is not a straightforward process, especially in terms of their development. However, Lim and Kor (2012) see values, which could be espoused and/or enacted, to have a role in affecting teachers' instructional practices. Other

N. A. Abdullah (✉)
Faculty of Education, The University of Hong Kong, Pokfulam, Hong Kong SAR, China
e-mail: azura.abdullah@ubd.edu.bn

F. K. S. Leung
The University of Hong Kong, Pokfulam, Hong Kong SAR, China
e-mail: frederickleung@hku.hk

researchers (Bishop 2007; Kadroon and Inprasitha 2013) believe that lesson study is a suitable way to study the teachers' values development in the classroom. Since the main highlight of the lesson study process is the collaborative work between teachers, perhaps through this active collaborative interaction, teachers' values can be made more obvious (Bishop and Seah 2008).

Brunei Darussalam has reformed its education system and nation-wide curriculum since 2009. It was introduced as *Sistem Pendidikan Negara abad 21* (also known as SPN21) or its English equivalent: National Education System for the 21st Century. The reformed mathematics curriculum puts emphasis on higher order mathematical thinking, holistic understanding of mathematical concepts and processes, and inquiry-oriented mathematics learning. In addition, the curriculum also puts stress on the students' mathematics skills on communicating their understanding, relating learned mathematics to real life problems and positive affective attitude to mathematics (Curriculum Development Department [CDD] 2010). This warrants for continuous professional development programs for school leaders and teachers to understand the reforms' objectives and contents. Among these programs, the Ministry of Education adopted a professional development approach originally from Japan called lesson study, specifically for the nation's centralized mathematics curriculum, which Stigler and Hiebert (1999) argued often helped in implementing reform ideas.

Lesson study is a teacher-initiated professional development approach set up in school and classroom settings. Stigler and Hiebert (1999) noted that the lesson study approach consisted of several successive actions plans. The first stage is for the teachers to investigate the lesson problem, plan the lesson, conduct the lesson, evaluate the lesson and reflect on it. The second stage is for the teachers to revise the lesson, conduct the revised lesson, evaluate the revised lesson and share the results. The pinnacle of lesson study is the collaborative work among teachers, and if needs be, outsiders, in conducting these action plans. Specifically, it is through the lesson study processes that "create changes in teachers' knowledge and beliefs, professional community, and teaching-learning resources" (Lewis et al. 2009, p. 286). It was anticipated that lesson study could help to remedy the gap between Brunei teachers' interpretation of the reform syllabus and their instructional practices in the classroom (Department of Planning, Development and Research [DPDR] 2014).

However, the transferability and adaptation of lesson study outside Japan have not always been successful. Fujii (2014) states that the practices of lesson study are often different from how the Japanese counterparts practice it. He believes that "the consideration of educational values is always tied to, influenced by, and reflected in, the key features of lesson study" (p. 78) and this emphasis on values may not always be present in the adaptation process. Perhaps teachers' values are not highlighted when Japanese lesson study is practiced elsewhere. Therefore, it is important to study the values aspect of lesson study in order to have an effective adoption of teachers' professional development approach in Brunei.

In order to have a holistic picture of teachers' espoused values, it is imperative to study values from three different processes in mathematics education (Tomlinson and Quinton 1986; Lim and Ernest 1997; Law et al. 2012): aims or intended outcomes; means or teaching/learning processes; and effects or actual outcomes (Tomlinson and

Quinton 1986, p. 3). Law et al. (2012) suggested these processes as pre-lesson, the observed lesson and post-lesson (p. 47). Meanwhile, Lim and Ernest (1997) suggested these processes in terms of levels in the curriculum, that is, the intended, implemented and attained curriculum (p. 37). In this study, values were explored from teachers' talk during lesson study processes of the planning stage (intended/aims), lesson teaching stage (implemented/means) and post-lesson meeting stage (attained/effects).

Raths et al. (1987) stated that people show signs of 'value indicators' when they go through a valuing process. They suggested that this process has seven stages of turning their beliefs into actions and cementing the final product as values. These stages include (1) choosing freely; (2) choosing from alternatives; (3) choosing after thoughtful consideration of the consequences of each alternative; (4) prizing and cherishing; (5) affirming; (6) acting upon choices; and (7) repeating (Raths et al. 1987, pp. 199–200). They go on to suggest that the seven stages can be further condensed into three main stages of choosing, prizing and acting, which are at the core of the process (p. 201). Therefore, it seems that these three core stages of the valuing process can be aligned with the lesson study process. Hereby, teachers' talk during planning sessions can be taken as value indicators at the choosing and prizing stages; teachers' talk during teaching sessions can be taken as value indicators at the affirming and acting stages; and teachers' talk during post-lesson discussion sessions can be value indicators at the repeating stage. Following Raths et al's framework, when value indicators have gone through all the stages above, only then the outcomes could be considered as values. In this study, teachers' values are being explored through their process of valuing in the lesson study processes of planning, teaching and post-lesson discussion.

Methodology

The research design is a case study using a qualitative approach (Yin 2003). This study attempted to study how teachers' values could be explored in the lesson study process by focusing on teachers' act of valuing their instructional practices. In this context, qualitative case study is useful in answering a "how" question "about a contemporary set of events over which a researcher has little or no control" (Yin 2003, p. 14). The study involved a school located in the capital of Brunei, Kelawar Primary School (pseudonym), that is considered as the pioneer of lesson study in mathematics due to its involvement at the early stage of a nation-wide lesson study project and teachers' familiarity with the lesson study approach. The lessons focused on were Year 4 mathematics lessons unit on addition and subtraction of fractions. There were six female mathematics teachers involved in the study with two teachers undertaking the actual teaching. One of the six teachers was only involved in the beginning of the study since she felt the need to remove herself from the lesson study due to her heavy commitment as she was also a Year 6 mathematics teacher.

The focus of the study was to explore the values and valuing process of a group of teachers involved in lesson study, especially the implementers, Teacher Melinda

Table: Teachers' participation in lesson study (pseudonyms were used to identify the teachers)

Teachers	Planning discussion		Addition of fractions lessons		Subtraction of fractions lessons		Post-lesson discussion
	1st	2nd	Class A	Class C	Class A	Class C	
Melinda	Yes	Yes	Yes**	Yes	Yes**	Yes	Yes
Ida	Yes	Yes	No	Yes**	Yes	Yes**	Yes
Athena	Yes	Yes	No	Yes	Yes	No	Yes
Carol	Yes	No	No	No	No	No	Yes
Rosanna	Yes	No	No	No	No	No	Yes
Alice	Yes	Yes	No	No	No	No	No
HM (head-mistress)	No	No	No	No	No	No	Yes

**Implementer

and Teacher Ida (pseudonyms were used to protect the identity of the teachers). The aim of the lesson study group was to develop and implement effective mathematics lessons on the unit "Fraction". Thus, the values explored in this study happened in natural settings and occurred without teachers specifically deliberating on the values.

The teachers were observed and video recorded during their planning discussion meetings, lesson implementation and post-lesson discussion meetings. Six teachers attended two planning discussion meetings (see Table). The teachers were a mix of lower and upper primary mathematics teachers. In the lesson planning sessions, two days' lessons on addition and subtraction of fractions were discussed. Four lessons, conducted in class A and C, in total were implemented and one post-lesson discussion was undertaken (see Table).

The handbook prepared by Peter Dudley (2014) was used as the tool to help teachers understand the lesson study process. Teachers found the handbook very useful and did adapt the ideas when they saw a need to do so. The forms they used were the lesson observation forms that aided them to focus on students' learning based on the lessons they planned. They also used post-lesson discussion forms to give their feedback on the lessons based on their observation and hence assessed the effectiveness of the lessons they observed.

As mentioned above, this study employs Bishop's (1988) idea of the socio-cultural dimension at pedagogical and individual levels. At these levels, the focus is on the teachers' valuing process and values and the choices they make in mathematics classrooms. Specifically, there are three flag points at this level where teachers' value indicators were explored, namely at the intended, implemented and attained points. Since teachers' values were explored at different flag points, for the sake of clarity, teachers' value indicators explored at the planning stage are called intended value indicators; teachers' value indicators explored at the teaching stage are called implemented value indicators; and teachers' value indicators explored at the post-lesson discussion stage are called attained value indicators. The exploration of teachers'

Table: Bishop's (1988, 2007) mathematical values and teacher's decisions on instructional practices

Component	Values	Prompts
Ideological	Rationalism	Teacher encourages students to give explanations, arguments or show mathematics proofs
	Objectism	Teacher encourages students to show or use diagrams, or concretising mathematics ideas
Sentimental	Control	Teacher encourages students to understand the process of routine calculations or check their answers and justify them
	Progress	Teacher encourages students to explore ideas beyond given examples
Sociological	Openness	Teacher encourages students to present and defend their ideas with whole class
	Mystery	Teacher encourages students explore their imagination on the wonder of mathematical ideas

values was done according to the "prompts" based on Bishop's (1988, 2007) categories as summarized in Table. All video recordings done at planning meeting, classroom teaching and post-lesson discussion stages were transcribed. Teachers' talk was studied to be value indicators of the prompts as shown in Table. Some of the talks were explicitly expressed showing clear indication of the prompts. For example, Melinda: "you are going to use this one (strip paper). It's either you fold or you draw the lines, the parts, and then you show me how you're going to get the answer". This is categorized as valuing objectism where the teacher encourages students to use concrete materials to show their mathematics ideas. However, some of the underlying values in teachers' talks were not clear and attempts were made to infer the prompts from excerpts of teachers' conversations or from teacher-students' interactions. For example, Ida: "Okay, why did I ask Albus to fold this into six?" Here the teacher used her questioning technique to seek possible reasons from students to Albus's answer. We inferred from this incident that Ida is encouraging her students to give explanations of the concept learned and this action seems to be valuing rationalism. More examples can be seen in the results and discussion section.

In chronological order from planning sessions, lessons implementations to post-lesson discussion meetings, thematic analysis (Braun and Clarke 2012) was done on the transcripts. Based on Bishop's framework Table, the teachers' talks were coded deductively according to the prompts they exhibited as value indicators to teaching fractions. At the same time, inductive thematic analysis was also used to describe the structural content of teachers' meetings and teachings.

Results and Discussion

The results of this study are presented in three sections. Section describes the value indicators of the group of teachers during their discussions in planning the lessons. Section reports on the flow of the lessons implemented and value indicators that the implementers enacted during their teaching. Section looks into the value indicators of the group of teachers during their post-lesson discussion.

Intended Value Indicators in Planning Sessions

According to Seah (2002), teachers reveal their values about mathematics and about the teaching of mathematics when they decide on the sequence and best strategies for teaching specific topics. Over two hours of recording was collected from the teachers' planning sessions in teaching a sub-unit of the topic fractions. The classes that they planned to teach were preceded by a few lessons on basic fractions before moving into the selected planned lessons of addition and subtraction of fractions. In these planning sessions, teachers' discussions seem to focus on three main themes: the structure of the lessons, the strategies of presenting the lessons, and the students' abilities to do the tasks set out for them. Teachers discussed the structure of the lesson in terms of recalling previous knowledge of fractions as introduction, developing anchor tasks and presenting the lesson development.

In the discussion on strategies to present the lesson, the teachers decided to present the topic using concrete materials and pictures or diagrams. The concrete materials they agreed on were coloured folding papers as bar models and pizza models. Teachers talked about students' abilities to 'see' the conversion of fraction and to correctly add two related fractions with the aid of diagrams and concrete materials. For the introduction problems and anchor tasks, teachers were planning on starting with simple problems as an introduction to addition of fractions and they decided to use diagrams first. They also voiced their concern on introducing the concepts abstractly in the first lesson, hence finalising their decision to use concrete materials in the lesson development followed by pictures or diagrams and then abstracts workings.

Later in the discussion on the lesson development, teachers were concerned with whether students are able to 'visualise' the process of getting equivalent fractions between related fractions of ½ and ¼. The teachers opted to use concrete materials, in this case coloured paper. The teachers planned to scaffold the sequence of the lesson to get students to 'see' the process of adding two related fractions by converting the fractions to their equivalent fractions and then adding them.

Since the teachers decided to conduct activities using concrete materials or pictures/diagrams, we infer that they were valuing objectism in the teaching of fractions. In their discussions, teachers used textbooks and teachers' guide as their point of reference when deciding on problems for students' activities. In the textbooks, fractions charts and bar models were used to depict fractions computations and, in the syllabus, it was recommended to "use concrete fractions models and fraction charts to help pupils add and subtract related fractions" (Curriculum Development Department 2008, p. 35). Perhaps, this may explain teachers' inclination to encourage students to use concretes or diagrams.

Implemented Value Indicators in Teaching Sessions

Two teachers implemented the four lessons, teaching two consecutive lessons each. The lessons were on addition and subtraction of fractions, a sub-unit under the topic of fractions. The teachers taught the lessons according to their plans as discussed. All four lessons had a similar structure with few differences in the teachers' nuances in lesson delivery. The lessons were structured in four phases as follows.

- Starting the lessons with objectives.
- Recalling previous knowledge.
- Solving problems.
- Summarising the lesson.

Lesson Objectives

Both teachers Melinda and Ida started their lessons by telling students the objectives of the lesson. These introductions were straight forward and do not show any signs of valuing mystery. In Melinda's case, she specifically wrote on the board "we are learning to…", stating the lesson objectives, and asked students to recite the objectives before starting the lessons. This is consistent with her pre-lesson study interview where she stated, "my introduction is always reading the learning intention and discussing the success criteria. Because you have to share before you go through the lessons". Perhaps from the teachers' actions, it could be inferred that they are valuing control, more so in teacher Melinda's practice, because they have established the targeted outcome for students to focus on. The actions also show that the teachers valued openness as they share with the students the aim of the lesson directly at the opening of the lesson.

Recalling Previous Knowledge

Both Melinda and Ida recalled their students' previous knowledge in the lessons as their lesson introductions. Ida used words such as *remember* and *recall* to draw students' attention on what they have learnt. She would ask students how to solve a problem and reminded them on how they have done it previously. She then proceeded to show the correct working. Ida tried to get her students to recall previously learnt procedural skills, followed by her demonstration of the skills and then encouraged her students to *show* the working referring to her example. Meanwhile, Melinda recalled students' previous knowledge before giving them a problem. She asked them conceptual questions such as definitions of fractions and procedural questions on adding fractions. She encouraged her students to *explain* their answers. Here, it could be inferred that Ida's actions valuing objectism whereas Melinda valuing rationalism.

Solving Problems

As planned, the teachers explained the concepts through students' activity of solving problems. During this section of the lessons, both teachers encouraged their students to use concrete materials such as fraction base, round paper pizza and 'strip paper' to solve the problems. This shows that the teachers are valuing objectism when they supported their students to use these and indeed other objects as well to concretise the mathematics concepts.

At the same time, both teachers also encouraged their students to explain their method of working by discussing it with their peers in pairs, or as whole class interaction. However, during the students and teachers classroom conversations, the students tended to give one word answers. When this occurred the teachers would resort to the concrete materials or pictures to get the students to *explain*. In this scenario, both teachers showed their attempts at getting students to discuss and explain their work with each other. This seems to imply that both teachers valued rationalism but were hindered by students' communication skills.

Even with students lacking in communication skills, both teachers encouraged their students to present and share their work in front of the whole class. By encouraging students to present their ideas to their peers, the teachers showed signs of valuing openness. To help better communication of ideas, the teachers encouraged their students to use their drawings as an aid to help with their explanations. This implies that the teachers valued objectism. This could be because in the Bruneian context, mathematics is taught in English, which is the second or even third language for most students. Hence it may be difficult for students to articulate their thinking to others and a medium in the form of concrete objects or diagrams was needed to communicate. In addition to this, both teachers 'took over' the students' presentation to further explain the steps of their workings to check their answers and emphasised the lesson development to the class. This action may indicate teachers are valuing control. Hence in this section, both teachers were inferred to be valuing objectism, rationalism, control and openness. These similarities may be due to their discussion of presenting problems to the students during planning.

Lesson Summary

Both teachers summarized their teaching towards the end of their lessons. Ida summarised both her lessons by asking the whole class the steps of adding or subtracting fractions. She wrote the steps on the whiteboard while explaining. Melinda only recapped her second lesson verbally with the whole class briefly at the end of her lesson. Both teachers resorted to verbal summary of their lessons.

Seah (2002) explained that teachers' values influenced them to make conscious decisions on their instructional practices. Based on the video recordings and transcripts, both teachers showed major hints in their actions of valuing objectism from their practice of folding papers and diagrams; and to a certain degree rationalism, control and openness in teaching the lessons.

Attained Value Indicators in the Post-lesson Session

Two hours of discussion were held after school with five of the teachers, with the headmistress present. The teachers watched clips of the classroom videos, looked at students' worksheets, and used their own classroom observation notes during this discussion. With the presence of the headmistress, the teachers' intentions and choices of teaching strategies were discussed. These talks provided more insights into the teachers' valuing process as to why they have chosen the instructional practices. The headmistress explained that she inquired the reasonings behind the strategies used by the teachers to understand the mathematics lesson as well as acting her role as an appraiser, to learn the teachers' expertise. However, during the presence of the headmistress, the teachers shared only the students' work and misconceptions. Further discussion on their teaching was done after the headmistress had left. It seems that with the presence of the headmistress, the teachers put the focus of discussion on students and not on themselves. This is perhaps because their headmistress is also their performance's appraiser.

Teachers referred to the students' work activity, with the aid of using diagrams, to assess students' understanding of the topic. From students' diagrams, teachers found three misconceptions in fractions where students used, first, different shapes to denote the same fractions. Second, different sizes of fractional shapes were used to denote same fractions. And third, students drew wrong diagrams but reached the correct answer. In addition to advocating the use of diagrams to assess students' thinking, the teachers reiterated the importance of diagrams to help students to *see* the concepts learnt. Consistently, teachers are valuing objectism. However, when focusing on the teachings, teachers highlighted their preference for moving the transition from concrete/diagram to abstract quickly as this would save time for students to find solutions especially during examination periods. This may depict teachers to valuing control.

Summary

Having a holistic picture of teachers' valuing process in the lesson study sessions provides insights into teachers' values. First, teachers' values could be explored from the value indicators in three different sessions of lesson study. This is especially true in Melinda's case, where she has shown to be valuing the importance of concretising the concepts of fractions and processes of adding and subtracting fractions. She consistently showed her persistence with the use of concrete objects and diagrams in her instructional practices during the planning discussions, her lesson presentations and the post-lesson discussions. We conclude that she valued objectism. However, we should take into consideration that these values may be dependent on the topic taught or the context of the lesson. In this case, the syllabus in Brunei recommended the use of concrete materials and diagrams in teaching fractions. It is important to note that in her lessons, she also attempted to get her students to *explain* to the whole class their ideas and justifying their answers. The responses were minimal, and Melinda encouraged her students to use concrete materials to aid their explanation. This could be due to the complex concept of fractions. Yusof and Malone (2003) found that fractions in Brunei primary schools is generally a difficult topic for students to learn. They concluded that this is due to two main reasons: lack of use of manipulatives in teachers' instructions of teaching fractions and lack of English proficiency from bilingual students. However, in this study, the teachers preferred to teach using concrete materials as they agreed to use a concrete-pictorial-abstract approach. Therefore, it could be the language difficulties that the students faced in expressing their ideas and thinking that impeded their communication skills. This might also be the reason why teachers showed an inclination towards valuing objectism more than rationalism.

Second, teachers' values could be understood further when the reasonings behind teachers' decisions of valuing a strategy or approach were examined. For example, teachers were valuing objectism by encouraging students to use concrete materials or diagrams for different purposes. In the planning sessions, the main reason is for students to 'see' the processes. In the teaching sessions, use of concrete materials or diagrams was to aid students in explaining their thoughts. In post-lesson discussions, it can be seen that teachers used students' drawings to assess their understanding of the lesson. Thus, as Fujii (2014) has stated, values could be reflected from and influenced by each of the lesson study processes. This is especially true in this case where values indicators reflecting objectism was consistently present at each process due to different reasons. In the planning sessions, objectism was valued to intend students to visualise the concepts; in the teaching sessions, objectism was valued to implement students' communication skills; and in the post-lesson discussions, objectism was valued to assess the attainment of the lessons.

Conclusion

There are two observations that we can take from this study. First, from the lesson study processes, we were able to look at value indicators at pre-lesson, during the lesson and post-lesson stages. The teachers' talk at these stages highlighted the teachers' preferences as they decided, implemented and pondered on the instructional practices, which can be treated as value indicators. Thus, as Fujii (2014) has stated, in the lesson study setting, the processes play a significant role in underlining teachers' values and these values are important for the effectiveness of lesson study as professional development for teachers in the mathematical learning of their students. However, this study only looked at the lesson study processes for one unit of lesson in a single cycle where no revised lessons were observed, limiting on how teachers' values may influence changes in their following lessons.

Second, by focusing on teachers' value indicators, their talks throw light on the reasons behind their preferences. With reference to Melinda's case, she has shown a pull between her valuing rationalism and valuing objectism to get her students to communicate their mathematical thinking. Due to the difficulties between the language of instruction and the mother tongue in a bilingual situation, and the challenges that come with it in the teaching and learning of mathematics, this study illustrates how the context may affect teachers' values and preferences in deciding their instructional practices in mathematics lessons. Teachers' values of mathematics teaching and learning might be different due to the bilingual context in the mathematics classroom of Brunei. Bishop (2001) explained that the mathematical values he elaborated were based on Western mathematics and may differ for different cultures. Perhaps, this study could elucidate teachers' valuing process and values in a bilingual culture.

Acknowledgements Part of this paper was taken from the first author's unpublished PhD research proposal that was submitted to the University of Hong Kong.

References

Bishop, A. (1988). *Mathematical enculturation: A cultural perspective on mathematics education*. Dordrecht: Kluwer Academic Publishers.

Bishop, A. J. (1999). Mathematics teaching and values education: An intersection in need of research. *ZDM Mathematics Education, 31*(1), 1–4.

Bishop, A. J. (2001). What values do you teach when you teach mathematics? *Teaching Children Mathematics, 7*(6), 346–349.

Bishop, A. (2007). Teachers' mathematical values for developing mathematical thinking through lesson study. *Processing of APEC-KHONKAEN International Symposium: Innovative Teaching Mathematics through Lesson Study (II)—Focusing on Mathematical Thinking*. CRME, Khon Kaen University.

Bishop, A. (2008). Mathematics teaching and values education: An intersection in need of research. In P. Clarkson & N. Presmeg (Eds.), *Critical issues in mathematics education: Major contributions of Alan Bishop* (pp. 231–239). New York: Springer.

Bishop, A. J., & Seah, W. T. (2008). Educating values: Possibilities and challenges through mathematics teaching. In M. H. Chau & T. Kerry (Eds.), *International perspectives on education* (pp. 118–138). London: Continuum.

Braun, V., & Clarke, V. (2012). Thematic analysis. In H. Cooper (Ed.), *Handbook of research methods in psychology: Vol. 2. Research designs* (pp. 57–71). Washington, DC: American Psychological Association.

Curriculum Development Department. (2008). *SPN21 syllabus year 4*. Brunei Darussalam: Ministry of Education.

Curriculum Development Department. (2010). *Framework and guidelines for curriculum and assessment: Mathematics year 6*. Brunei Darussalam: Ministry of Education.

Department of Planning, Development and Research. (2014). *Teachers and the teaching of Bahasa Melayu, English and Mathematics in primary schools in Brunei Darussalam.* Unpublished research report. Brunei Darussalam: Ministry of Education.

Dudley, P. (2014). *Lesson study: A handbook.* Retrieved from http://lessonstudy.co.uk/wp-content/uploads/(2012)/03/Lesson_Study_Handbook_-_011011-1.pdf.

Fujii, T. (2014). Implementing Japanese lesson study in foreign countries: Misconceptions revealed. *Mathematics Teacher Education and Development, 16*(1), 65–83.

Kadroon, T., & Inprasitha, M. (2013). Professional development of mathematics teachers with lesson study and open approach: The process for changing teachers values about teaching mathematics. *Psychology, 4*(2), 101–105.

Lewis, C., Perry, R., & Hurd, J. (2009). Improving mathematics instructions through lesson study: A theoretical model and North American case. *Journal of Mathematics Teacher Education, 12,* 285–304.

Law, H. Y., Wong, N. Y., & Lee, N. Y. L. (2012). A study of espoused values in Hong Kong's mathematics classrooms. *ZDM - The International Journal on Mathematics Education, 44*(1), 45–57.

Lim, C. S., & Ernest, P. (1997). *Values in mathematics education: What is planned and what is espoused?* Paper presented at the conference of the British Society for Research into Learning Mathematics, Nottingham, UK.

Lim, C. S., & Kor, L. K. (2012). 'Excellent' primary mathematics teachers' espoused and enacted values of effective lessons. *ZDM Mathematics Education, 44,* 59–69.

Macnab, D. S. (2000). Raising standards in mathematics education: Values, vision and TIMSS. *Educational Studies in Mathematics, 42,* 61–80.

Pa, N. A. N. (2009). *Nilai dan etika dalam pendidikan matematik* [Values and ethics in mathematics education]. Kuala Lumpur: Penerbit Universiti Malaya.

Raths, L. E., Harmin, M., & Simon, S. B. (1987). Selections from values and teaching. In P. F. Carbone (Ed.), *Value theory and education* (pp. 198–214). Malabar: Krieger.

Seah, W. T. (2002). Exploring teacher clarification of values relating to mathematics education. In C. Vale, J. Roumeliotis, & J. Horwood (Eds.), *Valuing mathematics in society* (pp. 93–104). Brunswick: Mathematical Association of Victoria.

Stigler, J. W., & Hiebert, J. (1999). *The teaching gap*. New York: Free Press.

Tomlinson, P., & Quinton, M. (Eds.). (1986). *Values across the curriculum.* London: Falmer Press.

Yin, R. K. (2003). *Case study research: Design and methods* (3rd ed.). London: Sage.

Yusof, J., & Malone, J. (2003). Mathematical errors in fractions: A case of Bruneian primary 5 pupils. In L. Bragg, C. Champbell, G. Herbert, & J. Mousley (Eds.), *Proceedings of the 26th Annual Conference of MERGA* (pp. 650–657).

15

The Role of Value Alignment in Levels of Engagement of Mathematics Learning

Penelope Kalogeropoulos and Philip Clarkson

Abstract Low levels of student engagement in mathematics education has been a growing concern in the Australian context and internationally. In this chapter, we will explore how value alignment strategies (Scaffolding, Balancing, Intervention and Refuge) could be used to rework conflict, resentment and disengagement of students in the mathematics classroom. When students are encouraged to discuss their individual thoughts and opinions, ideas and approaches, students' values and identities become more apparent. We also begin to explore the notion of identity as an extension to the completed work on value alignment strategies. We propose a possibility that value alignment strategies could be the catalyst in many more students proclaiming that 'mathematics is my most valuable subject!'

Keywords Values · Value alignment strategies · Student engagement · Student disengagement · Mathematical identity

Introduction

There are some things in life that appear logical such as 'don't play with fire' or 'never run with scissors in your hands'. Then there are other things that might seem equally natural to us as individuals, or as members of a family, but which we might not expect other people to agree with, such as 'voting in a referendum' or 'leaving our shoes at the door' (Rowan et al. 2007). In some contexts, decisions even around these seemingly personal decisions might be based on community driven ideologies. The things that we believe in, shape the way we view our world and the way we act within the world, through the decisions we make, indicate our values.

P. Kalogeropoulos (✉)
Faculty of Education, Clayton Campus, Monash University, Melbourne, VIC, Australia
e-mail: penelope.kalogeropoulos@monash.edu

P. Clarkson
Australian Catholic University, Locked Bag 4115, Fitzroy MDC, VIC 3065, Australia
e-mail: Philip.Clarkson@acu.edu.au

Values are abstract qualities that we recognise when we see them in action, through the decisions that people make, the way they react in a critical incident (Tripp 1993), and their engagement with any particular situation. Values move out of the abstract when we admire particular behaviours, attitudes and dispositions (Clarkson and Bishop 2000). There will always be differences in how people in a community interpret the same value, and the relationship between our values and life choices are not always transparent. Hence values, in many ways, are often implied rather than explicit. Therefore, it is not an easy process to identify with certainty, an individual's specific values. Raths et al. (1987) regard successful attainment of a value as involving all of seven criteria; choosing freely, choosing from alternatives, choosing after thoughtful consideration of the consequences of each element, prizing and cherishing a value, prizing a value through affirmation to others, acting on the choice of value made, and acting repeatedly to enact a value which gives rise to some pattern in life.

Values are acquired over time and that the negotiations of values between and amongst activity systems leads to values being challenged and refined. Due to this inevitable presence of competing and overriding values (Seah 2005), one value is not articulated in all situations. For the assessment and identification of valuing, some kind of triangulation is needed through the observation of multiple supporting activities (Seah 2018). This suggestion echoes (Raths et al. 1987) seventh criterion involving the rise of a pattern to life. In other words the realization of a value is always, necessarily a process of comparison (Graeber 2001).

Mathematical ideas develop everywhere even though people live in different cultures (Bishop 1988). We can find literally hundreds of different counting systems, using different symbols or no symbols, objects and materials varying with the cycles, or bases, used to deal with large numbers. Even in single countries like Papua New Guinea, a land of four million people but 800+ languages, there is a myriad of counting systems (Owens et al. 2018). The symbolic and religious properties of geometric figures are of more interest in some societies than others, as are the predictive powers of certain numerological practices, myths and ideologies around their symbolic importance.

Although many of these various ways of doing mathematics grew up in relatively isolated communities, today it is rare for communities to remain isolated. The movement of people across borders is taking place in unprecedented levels due to reasons such as armed conflicts, globalisation and regionalisation of trade and business (Seah and Andersson 2015). Thus, we are seeing growth in eclectic intersections of cultures that have not readily occurred in the past. Clearly this means different ways of doing mathematics are intersecting, and hence the values embedded in these different mathematic systems are also intersecting.

At quite a different level, any mathematics classroom can be seen as a community made up of diverse cultures and identity politics. Therefore, classroom interactions between mathematics teachers and students, and those among students themselves, are sometimes sites of contestations and conflicts. Values in mathematics education are inculcated through the nature of the mathematics studied and individual experience, and thus become the personal convictions that an individual regards as being important in the process of teaching and/or learning mathematics (Seah and Kalogeropoulos 2006).

Decisions and actions relating to the learning and teaching of mathematics in schools are linked to values amongst other aspects of the mathematics classroom. A student who values achievement, will study hard to pass an exam, and a student who values understanding perhaps will question theories until meaning has been constructed. Of course a student might value both achievement and understanding. Teachers also make decisions in mathematics classrooms depending on their values: For example, how often will the teacher provide a choice of activities in class, and how much routine practice is important? These teacher decisions could be embraced, simply accepted or maybe rejected by the students and this is usually indicated through student engagement. The student who values achievement may comply with the teacher's requests in order to receive a good grade, but the student who values understanding most may object or withdraw from the task if they do not see the value for furthering their understanding in it. How teachers respond in these situations can be useful indicators in turn to their valuing.

Mathematics classrooms are diverse learning environments. They always have been of course, but today such diversity is being recognized and in many classrooms at least tolerated if not celebrated. Hence if a teacher and students have moments when their views are not in agreement, this is now often dealt with openly. For example, a traditional-style teacher may value the automatic and rapid recall of number facts (e.g. single digit multiplication questions) and formulae and therefore teach mainly through closed-ended questions commonly found in mathematics textbooks. However her/his students may value relevance and communication more, hoping for inquiry-based tasks. In such a context the value alignment is at odds.

But it is not always like that. For example, a teacher who values group work in mathematics education will create experiences in his/her classroom for students to work together. However, a student who values independent work style may be troubled in these classroom situations. The way the teacher and the student react to this critical situation could also unveil another layer of values. The teacher may also value respect of preferred learning styles for each student, and therefore allow the student to work independently. On the other hand, the teacher may decide to hold onto this tension as they value disagreement as a fruitful space for learning (in this case, the child learning to work with others). The engagement of the student could be an indicator of whether their value has been compromised or neglected.

Since values and valuing are sociocultural in nature (Seah 2018) it is reasonable to argue that teachers and students cannot expect that the other party will 'always' share their valuing. However, in most classroom environments teachers and students will want to co-exist harmoniously and therefore they will adapt strategies to exhibit tolerance, respect and acceptance (general educational values and good character traits) without compromising their own values in mathematics learning.

From the above examples some very pertinent questions arise. How is harmony and engagement maintained in a mathematics classroom environment amidst the range of values present? How do teachers and students negotiate the differences that inevitably exist, so as to facilitate and maximise the learning of mathematics? If engagement is about inclusiveness, how do we help teachers perceive diversity as a positive? Let us assume a team teaching environment, which is a common practice in Australia. One teacher encourages problem solving in mathematics learning whilst the co-teacher prefers teacher modelling of mathematical procedures and quiet working environments. How do teachers negotiate such professional and pedagogical valuing conflicts?

Rather than trying to deal with all of these questions, this chapter focuses on strategies that four teachers have used to achieve value alignment with their students during critical incidents in mathematics learning. The value alignment strategies are part of the findings of the first author's recent PhD study conducted in Melbourne, Australia (Kalogeropoulos 2016). Values alignment strategies were observed to be used by the teachers to form a compromise, a negotiation or a change in mathematics learning for student engagement or re-engagement. Four such value alignment strategies were identified in the study and are perhaps one beginning to addressing the issue of students who show low levels of engagement in mathematics learning. The chapter then turns to an additional issue, that of mathematical identity, which also impacts on engagement in mathematics learning and requires consideration in value alignment.

The Four Value Alignment Strategies

There are a number of ways in which alignment of values between teacher and students can be achieved in the mathematics classroom. Kalogeropoulos (2016) used classroom observations particularly of critical incidents in the flow of teaching in lessons, student and teacher questionnaires, and interviews to collect data from four teachers and their 10–11 year old students in the same school in Melbourne, Australia. The data obtained was analysed using a grounded theory approach. From these analyses, four strategies of value alignment emerged: the scaffolding strategy, the balancing strategy, the intervention strategy and the refuge strategy. The four teachers consistently employed these four strategies during the three observed lessons (per teacher, hence 12 in total), when similar critical incidents arose.

The Scaffolding Strategy

The scaffolding strategy was adopted by the teachers observed when they came to their mathematics lesson with some type of preparation to scaffold the learning of the intended learning objectives. In one episode noted from the research observations, a teacher asked their 10/11 years old students to complete a challenging mathematics word problem independently. The set task included different types of information that seemed to confuse the students:

Ice cream task: Double scoops

Can you think of 7 different flavours?

You want to buy a double scoop with 2 different flavours.

Which different combinations could you choose?

You cannot repeat a combination.

The students attempted the task but soon complained about the difficulty of the task and their inability to solve it on their own, consequently disengaging from their mathematics learning. In an attempt to re-engage the learners in their mathematics learning, the teacher offered the students an enabling prompt (Sullivan 2018). Instead of 7 different flavours, the teacher modified the task for the students to think of 3 different flavours. The teacher also provided an option for the students to work with a peer for further support or scaffolding. In similar critical incidents, the four teachers were observed to frequently use these two approaches; value peer-support when they encouraged students to work together and to share ideas and propose solutions, and the teacher suggesting that solving a less complex task would help the student to understand the original more complicated task.

In order to maintain a functioning classroom environment amidst the range of values present, it helps when teachers and students understand one another's values and seek to bring them into alignment (Seah and Andersson 2015). This was achieved by the teachers when they scaffolded the task through an enabling prompt and/or offered peer support. On the other hand, tension was also maintained as the students helped each other to solve the word problem, allowing a fruitful space for learning through the use of a challenging task (Sullivan 2018). In both instances students and teacher are valuing task completion.

The teacher usually takes the leading role and uses her/his teaching craft in facilitating values alignment during classroom critical incidents (Tripp 1993). A shared vision (as seen in the example above) needs to be co-created, although in actuality the students could subscribe to these goals to different degrees. Thus, when a teacher is able to facilitate values alignment between what he/she values and what his/her students value, this promises to strengthen the relationship, and is one of the keys to nourishing teaching and learning practices (Seah and Andersson 2015) and ultimately ensuring students' engagement in mathematics learning. This is achieved through adopting new values for harmony but staying in the tension for learning growth.

The Balancing Strategy

The balancing strategy refers to a teacher accommodating student values that the teacher had not anticipated would be evident during the lesson. One example of this balancing strategy occurred during a critical incident that arose when students requested a calculator to check their answers in class. The crucial part of this incident unfolded as follows:

Student Can we use the calculator to check our answers?
Teacher No. You will not have a calculator during NAPLAN (state-wide testing) so you are just doing a disservice to yourselves.

In the first part of her response, the teacher refused to provide the students with a calculator. But to meet her value of accuracy, she decided to collect and correct the students' work at the end of the lesson, something she had not originally planned to do. Her response seemed to satisfy the demands of the students as they then continued to complete their work and submit it to the teacher for correction. The student's value of accuracy was not ignored by the teacher. In contrast, the value of accuracy was indeed accommodated by the teacher, suggesting that the teacher also shares this value with his/her students.

This example suggests that there can often be differences in how the same value can be interpreted and displayed in different ways (Rowan et al. 2007). The students valued accuracy by asking for a calculator to check their answers, but the teacher used an alternate approach to accommodate this value by collecting the workbooks for correction purposes. In this situation, the teacher has once again used her/his teaching craft in noticing the students' re-engagement with their mathematics learning, indicating that the value of accuracy has been negotiated and accepted in different terms.

It is worth noting that value alignment is not about facilitating a classroom situation in which everyone subscribes to the same interpretation of the value. This is a dynamic interaction when shifting positions from both parties is to be expected. In other cases, it could include the adoption of new, shared values. The deployment of a particular value alignment strategy depends on the situation, the learners and of course the values!

The Intervention Strategy

There were times in mathematics learning when the teachers were required to put their values aside and respond to the students' values that were being exhibited and required attention for student engagement. The extent to which a value is embraced and prioritised is always circumscribed by the lesson situation and hence responsive to the learning environment and the context of a conflict situation. For example, a student described their distress when they felt isolated and daunted during a particular mathematics lesson. The student was unable to complete their work and as a result their mathematical anxiety and emotions took over and left them feeling helpless and overwhelmed. The teacher intervened by offering the student one-on-one assistance. The intervention strategy used was to first provide immediate emotional support, closely followed by intensive teaching to help the student reengage with their mathematics learning. The teacher's humane values of care and compassion prevailed and when the student was eventually soothed, the mathematics learning was readdressed with a focus on understanding and success instead of the initial teacher value of independent work style (the student working independently to complete work). This example shows that students' stories and actions for learning mathematics can change as the contexts evolve (Seah and Andersson 2015).

Values have both cognitive and affective components. In this situation, the teacher has temporarily suspended what he/she initially valued in this context (independent work style) and allowed his/her overriding humane values (care and compassion) to deal with the situation and re-engage the student with their mathematics learning. Thus, valuing provides an individual with the will and determination to act in particular ways. In the above situation, values played a significant role in transforming a negative situation into a positive outcome in mathematics learning.

Refuge Strategy

The refuge strategy was a value alignment strategy used when the teacher put most (if not all) of their values to one side and used their authority in a manner that postpones their proposed lesson planning and instead focussed on the value orientations of the students. In this situation, the teachers found new values that aligned with their own and those of the students.

In one of the observed mathematics lessons, the students became 'stumped' by a problem-solving task that the teacher had planned. Even after the teacher attempted to explain the mathematical task numerous times, the students became agitated and disruptive. In an effort to reengage her students, the teacher made a spontaneous decision to play a mathematical game with the students. The chaotic classroom reformed to an enthusiastic environment as the teacher's and students' value of fun was embraced and aligned. Value alignment can therefore involve a teacher displaying flexibility and making detours from their intended lessons to accommodate new areas of interest (Kalogeropoulos and Bishop 2017).

Student engagement can therefore also be seen as an indicator for value alignment. As seen in this example, the teacher's prioritising of the valuing of fun was successful in maintaining classroom control. The teacher was conscious that the complexity of the given task was the trigger that lead to student disengagement and therefore used his/her professionalism to adopt a value alignment strategy in an attempt to reengage the learners with a mathematical game.

Classifying the Four Strategies

The four value alignment strategies described above were classified based on the extent to which the teachers retained their values after value negotiations had taken place with the students (see Fig.). They were deployed when a teacher began to notice signs of disengagement in her/his students. The teacher typically made small changes to the lesson when the scaffolding strategy was adopted and hence to their

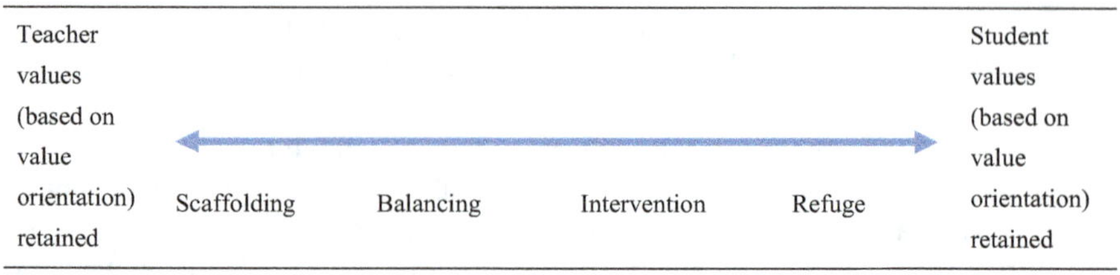

Fig. The four value alignment strategies

own values. For example, the task was simplified or students were encouraged to work with each other for support. In contrast, the refuge strategy at the other end of the continuum was used when disengagement was prevailing and the teacher needed to intervene, and hence re-prioritise her/his own values to engage the learners. For example, as was seen above, the teacher decided to suspend the lesson and play a mathematical game instead. Importantly these decisions were made "on the spot" during critical incidents and this was when it was clear that the teacher's craft and resourcefulness was used to attain values alignment in any classroom situation. The balancing and intervention strategies can be seen to lie between the two extreme strategies.

The extent to which a value is embraced and prioritised is responsive to one's environment and is thus not fixed (Seah and Andersson 2015). In the case of a classroom environment, when a teacher notices disengagement amongst the students, the teacher will assess her/his values and decide if they should be prioritised (continue with the planned lesson) or other values should override (adopting the above-mentioned value alignment strategies). In a rather paradoxical way, this adds to the extremely internalised and stable nature of values (Seah and Andersson 2015).

Teachers' values are expressed in their mathematics lessons, through their activities and discussions. As students enter these environments, they have their own set of values that may or may not be the same, similar, or different to the teacher's values. Attard (2011) suggested that the more powerful influence on engagement in mathematics for middle-year students appeared to be that of teachers. This influence can be viewed at two interconnected levels; the pedagogical repertoires employed by the teacher and the relationship that occurs between the teacher and the students. Analysing teacher pedagogies and student behaviours respectively through a "values lens" can provide us with insight as to why teachers make certain decisions in an attempt to reengage learners during mathematics learning. Hence in the case studies referred to above, the teachers were observed to act professionally and negotiate their values to form shared goals with their students (Seah and Andersson 2015).

Summary

The value alignment strategies support the notion that value priorities are dynamic and malleable although what is prioritized is in part contingent on the classroom context. What we value in the moment, also reflects our years of learning, influences from our historical experiences, and social interactions as members of the cultures to which we belong (Seah 2018). In the classroom, pedagogical activities provide interactions of what students, teachers and indirectly what the wider community value. Such interactions can expose what the students, the teachers and the community value similarly or differently. In effective classrooms, values are aligned or agreed upon by the different parties to maintain functioning activities in interaction. Therefore, values are acquired over time (in sociocultural contexts) and also challenged and refined on an ongoing basis, depending on interaction opportunities (Seah 2018).

During critical incidents within the flow of a lesson, a teacher often needs to choose amongst several alternatives. It is during this choosing activity, which may well be the employment of a particular value alignment strategy, the teacher's value priorities often become more clearly visible to an observer. Depending on the situation, teachers may be required to re-prioritise particular values. This requires teachers to firstly be aware of their values and personal convictions when the different values of teachers and students come together in interactions, resulting in value differences and value conflicts (Seah and Andersson 2015). When value alignment strategies are used effectively to reengage students in mathematics learning, teachers also demonstrate a capacity to acknowledge students' values, culture, knowledge, skills and dispositions, in an attempt to optimise and empower mathematical learning.

Mathematical Identity and Value Alignment

As can be seen from the above study pedagogical activities take place, in part, through the interactions of what teachers and students value. It is also clear that there can be misalignment between the teacher's values and those of the students, and indeed between students. The four teachers in the study utilised value alignment strategies and these sometimes introduced the co-creation of values that could be perceived as the agreed-upon, aligned values that facilitated the continued functioning of the activity systems in interaction (Seah 2018). However, is it too great a hope to expect value alignment will always be possible in the context of a mathematics classroom?

For example, when students are asked to complete a challenging problem, students may complain and begin to show signs of disengagement but a teacher's value of student perseverance may dominate and (s)he may decide to continue with their planned lesson, encouraging students to remain in the zone of confusion for a certain amount of time. In essence, the teacher is encouraging the 'complaining' students to accept and hopefully adopt the value of perseverance, similar to a parent encouraging a child to try a new food that they 'may' enjoy! In this instance, students' valuing can be and are being shaped in the mathematics education process with the teacher playing the role of value agent in mathematics teaching (Seah 2018).

Interestingly value alignment strategies may not always be so significant. For high performing students who really value achievement, the way mathematics is taught at school may have little impact on them., For them, their cultural values dominate, and possibly lead to decisions such as home tuition or enrichment classes to achieve the high performance that is desired. But in this instance, values, be they students, teacher, school or cultural, may not be the only consideration.

On reflecting on the results of the first author's doctoral thesis, and considering the preliminary statement to potential authors for this volume that in part asked that authors consider questions that would open up further lines of enquiry, we wondered whether the notion of identity (Cobb 2004; Lerman 2012; Sfard and Prusak 2005) may be another fruitful idea to explore concerning value alignment. Although both are formed through social interaction and developed over one's lifespan, identity represents an individual's subjective perspective of the self (Gatersleben, Murtagh and Abrahamse 2014). The extent to which a value is embraced and prioritised is responsive to one's environment and is thus not fixed (Seah and Andersson 2015). So can someone's mathematical identity influence their values in a mathematics classroom? If so, what implications does this have for the four value alignment strategies to successfully reengage students in mathematics learning, when there are other ongoing social, political and gender issues?

Others have indeed suggested a linkage between values and identity, and see values as an integral and indeed central aspect of identity (Gatersleben et al. 2014). Seah and Andersson (2015) propose that values are the convictions that an individual has internalised as being the things of importance and worth. Identity, however, is regarded as a broader concept that encompass many aspects of the self, including psychological processes (including behaviours) which people may adopt for maintaining and protecting the self (Breakwall 1986).

Identity may mediate the relationship between values and behaviour: a teacher who values understanding in mathematics learning will be motivated to plan lessons that support this value through the inclusion of challenging problems, real-life investigations and group discussion opportunities to discuss ideas and solutions. The identity of this teacher could be labelled as 'contemporary' and their value alignment strategies will probably differ to a teacher with a 'traditional style of teaching' identity.

However, the notion of identity is a contested one. Its fluidity of meanings makes it problematic when one is after clear guidelines that teachers might follow. And yet the fluidity of the identity notion mirrors that to some degree of the fluidity and problematic of the very notion of values and valuing: such fluidity can easily be seen when the various meanings of values and valuing used in the various chapters of this volume are compared. Hence seeking an interaction of the two notions, values and identity, may well be useful as they both impinge on the dynamic of the co-creation of teacher—students/students—students learning situations in mathematics classrooms.

Clearly the notion of identity does bring into play ideas of individuality, such a strong aspect of the western culture. But the counter surge to that in the classroom is a pressure to see all players as part of a community, which in some way has a shared identity; the classroom culture we strive for is not based on notions of the teacher and the students, or other multi-chotomies that can be envisaged, but a sense of a shared identity. Teachers and hopefully students are normally looking for ways so that there is trust, cooperation and support offered for those who are in need in the mathematics classroom context. Indeed that is the aim of the four value alignment strategies. But the ever- present individual identities that are in the classroom also bring into play the histories of all players, their ethnicities, their genders, their ideologies, their sexual orientations, and more (Chronaki 2016). Such diversity is brought into relief by noting the stereotypical images that are often portrayed in student aids such as textbooks (Clarkson 1993). Thus we see the interplay of values and identity, in all their meanings, as scope for broadening the questions that we ask.

Conclusion

Andersson et al. (2015) showed that changes in the contexts of teaching and learning can motivate students, even those who disengage either in the moment, or for longer periods of time, to productively re-engage with mathematics. As part of this process, value alignment strategies are employed by teachers to maintain a sense of harmony in a mathematics lesson and to help students develop a positive and active relationship with mathematics. Perhaps we must invest more of our time in helping students value mathematics by engaging their interest in the subject, helping them to identify their strengths and their weaknesses and focusing on making mathematics meaningful. More research is required in values, value alignment and alternative identity-work that pursues reconfigurations of mathematical subjectivity. The notion of dialogical mathematical education, where both teachers and students are required to reflect on and discuss their individual thoughts and opinions, ideas and approaches, embraces and acknowledges the mathematical identity of both parties. These interactions reveal values for consideration and value alignment may be the catalyst for facilitating meaningful mathematics learning.

References

Andersson, A., Valero, P., & Meaney, T. (2015). "I am [not always] a math hater": Shifting students' identity narratives in context. *Educational Studies in Mathematics, 90*(2), 143–161.

Attard, C. (2011). Engagement with mathematics: The influence of teachers. *Southeast Asian Mathematics Education Journal, 1*(1), 31–39.

Bishop, A. J. (1988). *Mathematical enculturation: A cultural perspective on mathematics education*. Dordrecht: Kluwer Academic.

Breakwall, G. M. (1986). *Coping with threatened identities London*. Methuen: Psychology Press.

Clarkson, P. (1993). Gender, ethnicity and textbooks. *Australian Mathematics Teacher, 49*(2), 20–22.

Clarkson, P. C., & Bishop, A. (2000). Values and mathematics. In A. Ahmed, J. Kraemer, & K. Williams (Eds.), *Cultural diversity in mathematics education* (pp. 239–244). Chichester, U.K.: Horwood Publishing.

Chronaki, C. E. (2016). Mathematics education as a matter of identity. In M. A. Peters (Ed.), *Encyclopedia of educational philosophy and theory* (pp. 1–6). Singapore: Springer Nature Singapore.

Cobb, P. (2004). Mathematics, literacies and identity. *Research Quarterly, 39*(3), 333–337.

Gatersleben, B., Murtagh, N., & Abrahamse, W. (2014). Values, identity and pro-emvironmental behaviour. *Contemporary Social Scinece, 9*(4), 374–392.

Graeber, D. (2001). *Toward an anthropological theory of value: The false coin of our own dreams.* NY: Springer.

Kalogeropoulos, P. (2016). *The role of value alignment in engagement and (dis)engagement in mathematics learning.* (Unpublished doctoral dissertation). Monash University, Melbourne.

Kalogeropoulos, P., & Bishop, A. J. (2017). What is the role of value alignment in engaging mathematics learners? In A. Chronaki (Ed.), *Proceedings of the 9th International Mathematics Education and Society Conference* (Vol. 2, pp. 603–610). Volos, Greece: Mathematics Education and Society.

Lerman, S. (2012). 'Identity' as a unit of analysis in researching and teaching mathematics. In H. Daniels (Ed.), *Vygotsky and sociology* (pp. 175–191). London: Routledge.

Owens, K., Lean, G., Paraide, P., & Muke, C. (2018). *History of number: Evidence from Papua New Guinea and Oceania.* New York: Springer.

Raths, L. E., Harmin, M., & Simon, S. B. (1987). Selections from 'values and teaching'. In J. P. F. Carbone (Ed.), *Value theory and education* (pp. 198–214). Malabar, FL: Robert E. Kreiger.

Rowan, L., Gauld, J., Cole-Adams, J., & Connolly, A. (2007). *Teaching values.* Marrickville, N.S.W.: Primary English Teaching Association.

Seah, W. T. (2005). Immigrant mathematics teachers' negotiation of differences in norms: The role of values. In M. Goos, C. Kanes, & R. Brown (Eds.), *Mathematics education and society: Proceedings of the 4th International Mathematics Education and Society Conference* (pp. 279–289). Brisbane: Griffith University.

Seah, W. T. (2018). Improving mathematics pedagogy through student/teacher valuing: Lessons from five continents. In G. Kaiser, H. Forgasz, M. Graven, A. Kuzniak, E. Simmt, & B. Xu (Eds.), *Invited lecturers from the 13th International Congress on Mathematical Education* (pp. 561–580). Cham, Switzerland: Springer.

Seah, W. T., & Andersson, A. (2015). Valuing diversity in mathematics pedagogy through the volitional nature and alignment of values. In A. Bishop, H. Tan, & T. Barkatsas (Eds.), *Diversity in mathematics education* (pp. 167–183). Cham, Switzerland: Springer.

Seah, W. T., & Kalogeropoulos, P. (2006). Teachers of effective mathematics lessons: What gets valued in their practice? In J. Ocean, C. Walta, M. Breed, J. Virgona, & J. Horwood (Eds.), *Mathematics: The way forward* (pp. 279–287). Melbourne, Vic.: The Mathematical Association of Victoria.

Sfard, A., & Prusak, A. (2005). Telling identities: In search of an analytic tool for investigating learning as a culturally shaped activity. *Educational Researcher, 34*(4), 14–22.

Sullivan, P. (2018). *Challenging mathematical tasks.* Oxford: Melbourne, Victoria.

Tripp, D. (1993). *Critical incidents in teaching: The development of professional judgement.* London: Routledge.

Permissions

All chapters in this book were first published in VVME, by Springer Nature; hereby published with permission under the Creative Commons Attribution License or equivalent. Every chapter published in this book has been scrutinized by our experts. Their significance has been extensively debated. The topics covered herein carry significant findings which will fuel the growth of the discipline. They may even be implemented as practical applications or may be referred to as a beginning point for another development.

The contributors of this book come from diverse backgrounds, making this book a truly international effort. This book will bring forth new frontiers with its revolutionizing research information and detailed analysis of the nascent developments around the world.

We would like to thank all the contributing authors for lending their expertise to make the book truly unique. They have played a crucial role in the development of this book. Without their invaluable contributions this book wouldn't have been possible. They have made vital efforts to compile up to date information on the varied aspects of this subject to make this book a valuable addition to the collection of many professionals and students.

This book was conceptualized with the vision of imparting up-to-date information and advanced data in this field. To ensure the same, a matchless editorial board was set up. Every individual on the board went through rigorous rounds of assessment to prove their worth. After which they invested a large part of their time researching and compiling the most relevant data for our readers.

The editorial board has been involved in producing this book since its inception. They have spent rigorous hours researching and exploring the diverse topics which have resulted in the successful publishing of this book. They have passed on their knowledge of decades through this book. To expedite this challenging task, the publisher supported the team at every step. A small team of assistant editors was also appointed to further simplify the editing procedure and attain best results for the readers.

Apart from the editorial board, the designing team has also invested a significant amount of their time in understanding the subject and creating the most relevant covers. They scrutinized every image to scout for the most suitable representation of the subject and create an appropriate cover for the book.

The publishing team has been an ardent support to the editorial, designing and production team. Their endless efforts to recruit the best for this project, has resulted in the accomplishment of this book. They are a veteran in the field of academics and their pool of knowledge is as vast as their experience in printing. Their expertise and guidance has proved useful at every step. Their uncompromising quality standards have made this book an exceptional effort. Their encouragement from time to time has been an inspiration for everyone.

The publisher and the editorial board hope that this book will prove to be a valuable piece of knowledge for researchers, students, practitioners and scholars across the globe.

List of Contributors

Monica E. Carr
Melbourne Graduate School of Education, The University of Melbourne, Melbourne, VIC, Australia

Julia Hill, Jodie Hunter and Roberta Hunter
Institute of Education, Massey University Albany, Auckland, New Zealand

Yüksel Dede
Gazi Education Faculty, Department of Mathematics and Science Education, Gazi University, Ankara, Turkey

Qiaoping Zhang
The Education University of Hong Kong, Tai Po, Hong Kong SAR, China

Yip-Cheung Chan
Department of Curriculum and Instruction, The Chinese University of Hong Kong, Shatin, Hong Kong

Ngai-Ying Wong
The Education University of Hong Kong, Tai Po, Hong Kong

Wee Tiong Seah
The University of Melbourne, Melbourne, VIC, Australia

JeongSuk Pang
Korea National University of Education, Cheongju, South Korea

Nagisa Nakawa
Kanto Gakuin University, 1-50-1, East Mutsuura, Kanazawa Ward, Yokohama, Kanagawa 236-8501, Japan

Philip Clarkson
School of Education, Australian Catholic University, Fitzroy MDC, VIC 3065, Australia

Takuya Baba
Hiroshima University, 1-5-1, Kagamiyama, Higashi-Hiroshima, Hiroshima 739-8529, Japan

Isao Shimada
Nippon Sport Science University, Setagaya, Japan

Douglas Lyman Corey
Brigham Young University, Provo, UT 84602, USA

Hiroyuki Ninomiya
Saitama University, Saitama, Japan

Ernest Kofi Davis
School of Educational Development and Outreach, College of Education Studies, University of Cape Coast, Cape Coast, Ghana

Ernest Ampadu
University of Ghana, Accra, Ghana

Annica Andersson
University of South-Eastern Norway (USN), 3199 Borre, Norway

Lisa Österling
Stockholm University, Stockholm, Sweden

Esra Selcen Yakıcı-Topbaş Yüksel Dede and Yüksel Dede
Gazi University, Ankara, Turkey

Fatma Nur Aktaş
Division of Mathematics Education, Gazi Education Faculty, Department of Mathematics and Science Education, Gazi University, Ankara, Turkey

Nor Azura Abdullah
Faculty of Education, The University of Hong Kong, Pokfulam, Hong Kong SAR, China

Frederick Koon Shing Leung
The University of Hong Kong, Pokfulam, Hong Kong SAR, China

Penelope Kalogeropoulos
Faculty of Education, Clayton Campus, Monash University, Melbourne, VIC, Australia

Philip Clarkson
Australian Catholic University, Fitzroy MDC, VIC 3065, Australia

Index

www.ingramcontent.com/pod-product-compliance
Lightning Source LLC
LaVergne TN
LVHW051045220926
848485LV00001BA/14
9781639871803